REGION AS A SOCIO-ENVIRONMENTAL SYSTEM
An Introduction to a Systemic Regional Geography

The GeoJournal Library

Volume 16

The titles published in this series are listed at the end of this volume.

Region as
a Socio-environmental System

An Introduction to
a Systemic Regional Geography

by

Dov Nir
Leon J. and Alice K. Ell Professor
of Environmental Studies,
Hebrew University, Jerusalem

KLUWER ACADEMIC PUBLISHERS
DORDRECHT / BOSTON / LONDON

Library of Congress Cataloging-in-Publication Data

Nir, Dov.
    Region as a socio-environmental system: an introduction to a
  systemic regional geography / Dov Nir.
       p.   cm. — (GeoJournal library ; 16)
  Includes bibliographical references.
  ISBN 0-7923-0516-7
  1. Geography. I. Title. II. Series.
  G116.N57   1990
  910—dc20   89-20099

ISBN 0-7923-0516-7

Published by Kluwer Academic Publishers,
P.O. Box 17, 3300 AA Dordrecht, The Netherlands.

Kluwer Academic Publishers incorporates
the publishing programmes of
D. Reidel, Martinus Nijhoff, Dr W. Junk and MTP Press.

Sold and distributed in the U.S.A. and Canada
by Kluwer Academic Publishers,
101 Philip Drive, Norwell, MA 02061, U.S.A.

In all other countries, sold and distributed
by Kluwer Academic Publishers Group,
P.O. Box 322, 3300 AH Dordrecht, The Netherlands.

printed on acid free paper

Printed in the Netherlands

CONTENTS

FOREWORD: AUTHOR'S CREDO

... The ideal society is a point on a receding horizon. We move steadily towards it but can never reach it.

Herbert Read

This text is both introverted and extroverted: it is intended for fellow geographers, many of whom are questioning the future of geography, and for non-geographers, whose perception of geography has been misguided by their primary school experience of geography presented as an inventory of topographical objects. The main aim of this text is to stress the socio-cultural value of regional geography. I believe that a good regional geography contributes to a better relationship between man and society and between man and his environmental challenges; geographical knowledge can contribute to a deeper understanding of the mutual relationship between different components of our existence. Today, regions are the framework of studies not of geographers only, but also of administrators, sociologists, economists, planners and politicians. I hope that this text will serve them as well.

This is a very personal text, conceived after more than thirty-five years of teaching and researching in geography at the Hebrew University of Jerusalem. I ask myself, what has been my contribution to the society in which I live; what social contribution have I made towards a *better understanding between people*.

Every one of us is a part of society and each of us is a unique creature, having a unique personality, existing only once in all of time. The commitment of a human being is, therefore, two-faced: as a member of society, man is constrained by its laws and limits, and, at the same time, he has the responsibility to himself as an individual not to waste the unique chance of life that is given to him. Life should be an effort to harmonize the relationship between 'homo socialis' and 'homo individualis'. The relationship between the individual and society seems to me to be the basic problem of human existence. Achieving harmony between them is, perhaps, the goal of human existence.

The way to seek after this goal is through education. Education is not brainwashing, it is not indoctrination, it is not memorization; education is the inner persuasion of a human being to commit himself to an aim: in the case stated, to live in harmony with society. But with what society—family, neighbourhood, city, nation, humanity as a whole? To achieve any of these seems an impossible task. There are such great differences between people, there is so much antagonism within societies, there are so many oppressors and oppressed, so many demagogues among the decision-makers, that it seems a goal beyond reach. The ideal society would be a free, anarchic society [an – without, archos – ruler], a society in which each individual is responsible for the relationship between himself and the society. By inner persuasion, we must live by making the maximum contribution of our physical and mental assets combined with minimal charge against and exploitation of the society. We must contribute to society as much as possible because, directly and indirectly, we enjoy the contributions of the global society in which we live and of which we are a part.To achieve this goal, we must know not only ourselves but also the society in which we live.

A society is not uniform. It is composed of mosaics of people of varying characteristics, structured in different patterns and groups, the qualities of which we must know because upon them depends our own place in the society. Were the world uniform of feature and society, there would no place for regional geography. But because the world varies in form and its societies are different, regional geography as the *study of the differentiation of the world's surface and the people who live on it* is an important tool for understanding the society in which we live, particularly when our goal is to live with it in harmony.

To achieve harmony does not mean an obligation to accept, willy-nilly, all the qualities found in society. Any society is likely to contain segments professing ideas, perceptions and iconographies other than certain of ours which we are unwilling or unable to renounce. To live with others by respecting their ideas but not renounce our own is the way to achieve harmony within a society. What I have in mind is, of course, the pluralistic society.

This goal of living in harmony within a pluralistic society, according to personal expectations and ideals, seems remote. The anarchic – not an anarchistic! – ideal has been perceived as a horizon, ever receding as we advance towards it, as something unattainable. Gershon Sholem attests [1982] that although he sympathized in his youth with Tolstoian anarchism, he could not believe the human mind capable of achieving the anarchic ideal.

But I do not despair. Other ideals, also, which today have been realized to some significant degree, only a few generations ago seemed just as incapable of attainment. Six generations ago, it would have been impossible for a Ukrainian muzik, a slave in fact, to imagine that one of his descendents might take the place of the omnipotent czar and reside in the Kremlin. But it happened, less than a hundred and fifty years after the abolishment of slavery in Russia. Could a black slave in the American South of the early 19th century have imagined that one of his descendents would seek the nomination of a major political party as its candidate for the U.S. presidency? This too happened, in 1984, and the event was considered within the bounds of normal procedure. The ideal of an anarchic-pluralistic society, where the individual and society will live in harmony, is no more fantastic an imagining than these examples were for their time; it is not a receding horizon but a steady goal for coming generations.

What can regional geography contribute to this goal? Regional geography deals with the *challenges* posed to a certain *society* at a certain *place* on the globe and with the *responses made by that society*. Its focus is the study of differentiation between societies, and this is the basic subject matter for the study of pluralistic societies. A central human value is freedom, which may be defined as *the right to be different one from another*. By studying differences, then, we stress the importance of freedom. Rather than a perspective of world-wide uniformism, whether cultural, social or political, I propose the perspective of a pluralistic world. This perception is, to my view, the main contribution of regional geography to humanity.

The living mind and living science are in perpetual renaissance. But because each healthy and creative generation produces its own answers to its particular problems, there is no such thing as a 'new' science or 'new' geography. A 'new' geography, a 'new' cinema is obsolete with the next wave bearing another 'new' geography or 'new' cinema. Human thought, as the Phoenix, lives through constant renovation.

Four basic positions have been proposed concerning the meeting of new ideas and new methods with existing ones [Reynaud, 1974]. These are formalized into a matrix: accepting neither new ideas nor new methods; rejecting new ideas, but accepting new methods; accepting new ideas, but not new methods; and, finally, accepting both new ideas and new methods.

The first situation is total negation of everything new, be it method or idea, a fundamentalism which does not take into account any change produced by time. Although this fundamentalist approach has been the basic approach of ultra-orthodox religions, whose main interest has been to perpetuate certain dogmas and ways of life, it cannot be the way of a living thought and a living science. This way of thinking has already exacted from humanity too great a toll.

The last position is the opposite of the first; it is total revolution and nihilism, throwing out all that has been achieved, discarding ante-revolutionary ideas and methods. But if everything before was so wrong, how did we arrive at these new right ideas? This approach of negation of all 'old' ideas and methods is unjustified and, to me, unacceptable.

Accepting new ideas but continuing to treat them by existing methods seems to be a waste of the spirit because, lacking tools appropriate to the subject-matter, one's effort will not be rewarded. Imagine handling vast amounts of information using only a sheet of paper and a pencil? On the other hand, the third alternative, using new methods for treating obsolete ideas, is a waste of mind, time and material, because even the most elaborate and elegant treatment, if not seconded by new ideas, is incapable of creativity or innovation.

This matrix presenting the meeting of old and new approaches does not exhaust the problem. If we reject the two extreme positions – uncompromising fundamentalism and nihilistic revolution – and if we accept the proposition that each generation has the *right and duty* to create new methods and new ideas, then we must find a way of conciliating the second and the third situations in the matrix, so that pre-existing ideas and methods which are still relevant as new situations arise can be either preserved or modified in order to contribute to the ideas of the present generation. My attitude in this work is not to throw away ideas simply because somebody expressed them before our time, but to weigh and judge each idea, each paradigm, each model – however the creations of the human mind may be labelled – in their relationship to the problems of a new generation. It is this generation which must live with the actual problems, and which must seek, in self-justification, an answer to the moral, material or political questions of its existence. It is unthinkable that a generation should live only according to ideas created by past generations; on the other hand, ideas of past generations which the present generation considers vital for itself should be integrated in its iconography.

I propose in this text to examine the idea of regional geography, produced by generations of geographers, to weigh it and to judge it, and to model it in relation to the actual world and the present problems of geography.

This is not an objective book. From the beginning it was written in the profound persuasion that systemic regional geography can and should be the center and the essence of geographical research.

These ideas germinated for years and were stimulated by suggestions and discussions with many colleagues. Their first budding forth was my participation in the annual convention of the Association of Israeli Geographers in Haifa, 1978, where I spoke on the 'totalitarianism of great numbers' and on the rehabilitation of 'noise'. These ideas were later published [Nir, 1985]. My sabbatical leave of 1984/85 gave me the opportunity to use the libraries of the Universities of Califiornia at Berkeley and Maryland at College Park, and the Library of Congress, as well as to meet Profs. H. Steinberg, J. Parsons, K. Corey, R. Harper and H. Brodsky, with whom I discussed some of the ideas in this text. On rather short visits to Paris [February 1986] and London [February 1987] I met Profs. P. Claval and D. Diamond, who were very helpful to my work; my sincere thanks are due also to the Librarian of the Royal Geographical Society. I am very much indebted to Dr. S. Skolnikov, from the Institute of History and Philosophy, Hebrew University, who read chapter six, and to Mrs. Sabina Schweid, from the Department of Education, Hebrew University, who read part of chapter four. With my Maître Jean Gottmann I discussed many parts of this text. To all of them I wish to express my deep appreciation.

Jerusalem, July 1, 1987

THE STATE OF THE ART

Geography remains in a state of methodological turmoil... Geographers have reduced not only where they study and what they specialise in studying, but also how they study, to its furthest elementary fragments. The situation is becoming untenable.

Martin J. Haigh, 1985, p. 201

A look at the geographical literature over recent years will show you, perhaps to your surprise, that the prophecies of doom for regional geography have been discredited. Regional geography is not dying. On the contrary, there are signs of new growth. A glance at the papers quoted in the 1984 April-June issues of Current Geographical Publications reveals that between 38% and 49% of the listed publications have been classified as regional. Of course, classification depends on the method used, and perhaps in other bibliographical reviews the percentages would be different. Nevertheless, these figures evidence a renewed interest in regional geography. This state of the art contradicts the attitude of the ruling geographical establishment, by which I mean the University departments, where regional geography is nearly ignored.

Most of the prevailing opinions on regional geography are biased, the result of prejudice of one kind or another. Further, the classical regional geography of the beginning of this century has been interpreted inaccurately. Last but not least, regional geography has the drawback of being a difficult and complex subject-matter whose mastery requires a lengthy apprenticeship. The negative appraisals do not lack a certain element of antagonism between generations, as each generation considers its duty to be, first of all, the negation of the achievements of previous generations, so as to be able to forward its particular message. '...In order to live and create, we must destroy the tradition from which we stem' [Olsson, 1979]. Despite this atmosphere – or because of it – I propose to introduce into the current discussion new approaches to regional geography, the most advanced of which, and that which offers the best promise of taking geography out of its current ideological stalemate, is the systems approach.

GEOGRAPHY, AS LIFE ITSELF, IS BUILT ON DICHOTOMIES

The dilemmas of geography can be seen at least partly as issuing from the turmoil that is felt, as well, in other branches of contemporary thought. In both arts and sciences it seems that everything is undergoing a reappraisal. Our generation is witness to the greatest material change in history. Never before have communications, transport, exchange of information, diffusion of goods and ideas been so rapid and so world-wide. These conditions of knowlege diffusion reach into every activity of mankind, and geography cannot be excluded from the general trends. The rapid spread of innovation has brought about another trend, common again to most of the arts and sciences: the fast-paced succession of generations in the various disciplines. What has been achieved through years of labour is made obsolete in a short time by the achievement of someone else, and yesterday's innovations today have already become outdated. If we look at how rapidly new paradigms emerge only to be replaced even more swiftly by still newer ones [Jones, 1985], we see that in the two last generations more new ideas, paradigms and models have been conceived than in the hundreds of years which preceded, even if some of the new ideas are only old ones revived and modified to present-day trends of thought.

The turnover in generations is quite characteristic of geography. Perhaps the most intriguing phenomenon is that most of the innovators in this field have been young people, novices in fact, who were dissatisfied with the way taken by their teachers and expressed their innovating ideas in their Ph.D. theses, at the beginning of their academic careers [Bunge, 1960; Harvey, 1969; Guelke, 1974; Gould, 1977]. In the past, one would have expected a basic epistemological-philosophical text to be a work achieved after years of experience in the field; but, of course, this is not the way of revolutionaries.

Then, after some years, another phenomenon comes to the fore: after the 'Sturm und Drang' period, after a little more experience has been acquired, retrospection and a certain readiness to compromise set in as one's horizons expand, a current of reformism enters the mind [Berry, 1973; Relph, 1977; Guelke, 1981], and the radical consequences of youth seem to require some slowing down. This process is quite well known by psychologists of education, the process of 'home – out of home – back to home'.

Geography was not exempted from these trends of antagonism between generations, as they were only a faster and more vehement process of the normal evolution in a creative society. But this alone cannot explain the explosion of geographical thought in every possible azimuth. Fertile ground for this intellectual explosion and the cause of this development may be found in the very nature of geography, which is *an intellectual occupation of mind, built on dichotomies.*

Whereas dichotomies are dangerous to disciplines, they are, in the main, the most creative background for the human mind. Human existence is but a series of dichotomies. The basic dichotomy, Life and Death, is a fundamental element of our existence; no human being is excluded from its problematics. To live knowing that each day may be one's last is, certainly, not an easy way to live, but it is inescapable.

Another universal dichotomy is that of psyche and corpus, or mind and body, the integral parts of the human being. Each of them exacts sacrifice and devotion; each of them makes an effort to dominate life; life is, in most cases, a compromise between these two diagonally opposed parts. We could continue ad infinitum with the dichotomies which make up life: materialism/idealism, quality/quantity, measurable/inmeasurable, and so on. *Life is but an unending effort to overcome dichotomies and to try to live within them,* by seeking possibilities of reconciliation, of bridging between opposites seemingly unbridgeable. In some cases, we have a certain success: medical treatment can delay, sometimes, the coming of death; by taming the self, we can reconcile mind and flesh; a middle path can be found between our materialist and idealist visions.

To overcome the problematics of dichotomies, I propose that the elements of the dichotomy be considered not as separate structures, but as *extreme parts of the same system*: in this example, of the human being. What is essential is the functioning of the human being, of the system, within and in spite of its many dichotomies. Each human act is the result of the combined functioning, with or without compromise, of both parts of the dichotomy. This leads to the understanding that most dichotomies are *apparent* dichotomies only, that they in fact represent, as stated above, only extreme positions in a single system.

Thus, if geography is an intellectual occupation built on dichotomies, the presence of dichotomy should not, in and of itself, be cause for crisis, dissatisfaction, or negation. What must be done is to answer the dichotomies. The two basic dichotomies of geography today are the dichotomy man/environment – the basis of the discussion in geography as a social or an environmental science – and the dichotomy place/space, which grounds geography as a science of spatial differentiations or of spatial processes. These are, in other words, at the core of the rift between regional and topical geography. As geography deals with the mind, one might ponder another dichotomy: whether geography is art or science. The dichotomy objectivism/subjectivism, another controversial issue methodologically and epistemologically, is not endemic to geography alone, but is nevertheless an issue which also should be examined.

The first of these dichotomies, man/environment, is the basic one distinguishing geography from other disciplines; the second one is basic to the inner structure of geography; the last two are common to some other disciplines in the social sciences which are also engaged in the current vogue for epistemological self-examination.

The dichotomy man/environment

This is the basic dichotomy of geography. Depending on one's point of view, it is the basic strength or weakness of geography. In past generations, most appreciations of geography were attributed to this 'virtue', whereas modern critics see in this dichotomy the Achilles' heel of geography. To the latter it seems impossible, either epistemologically or methodologically, to adequately treat a subject whose aspect includes elements of both human and natural disciplines [Kimble,1951].

The divergence of views is traceable to the different opinions on the place of geography within a disciplinary framework. To Vidal de la Blache [1913], geography deserves a special niche in the natural sciences: '....within the group of natural sciences, among which it undoubtedly belongs, [geography] holds a place apart'[author's translation]. Half a century later [Fielding, 1974], most of the human geographers had come to define geography as a a social science.

Previous generations held the position of geography to be the logical outcome of the intellectual effort to master the scholastic approach to life phenomena, in which everything should be ordered and separated into different categories called disciplines. The division of Nature into spirit and body, which seems so natural to Judaeo-Christian civilisation, is not universal; in Buddhism, Hinduism, and Shintoism such a differentiation does not exist, and soul is attributed to a tree, a stone, or a waterfall [James, 1967]. But the current trend in geography* is to eliminate every aspect of physical geography from its research and teaching [Gregory, 1978]. This trend is perhaps nourished, subconsciously, by the fear of neo-determinism being introduced into geography. Nothing could be further from my purpose than the intention to introduce determinism into geographical thought; neither do I intend to treat in this text the history of determinism, possibilism and probabilism in geography, as it has been discussed exhaustively by Gregory [1978; Johnston, 1983]. Through technology, capital, organisation and motivation [Gottmann, 1957], most natural challenges and constraints can be overcome, tamed and turned to good use. Nevertheless, natural challenges exist, and it is only due to man's ability, efforts, motivation and perception that they are not more determinative of the geographical reality. To deny the existence of environmental challenges – I use the term 'challenges' not incidentally, but in the persuasion that man's existence is constantly challenged by natural, social and political forces – is to endanger man's existence on Earth.

Even today, in an epoch when Man by his technological sophistication can withstand immense natural challenges and hazards, he acts not in a vacuum but in a certain environment. The most developed technological society is not free of floods, droughts, storms, and other such phenomena; all these influence, although perhaps only sporadically or temporarily, his daily economy and life. Consider floods, for example, which can be an acute problem even in the most developed of countries. One need only read American newspapers in the spring to perceive the dimensions of this problem in fatalities and economic loss, not to speak of less technologically developed countries, such as India and Pakistan, where periodic floods are even more devastating. One cannot deal with social structures without treating the basic problems of their very existence. Society and its physical environment is not a dichotomy: each is part of a whole, a *system*.

Perhaps I am anticipating somewhat the central thesis of this book – the region considered as a system – by arguing that natural challenges are not a subject per se in geography, but a part of the regional system, together with the social, cultural, economic and political challenges of a region. The example of floods illustrates this approach quite well. To respond to this natural challenge, one must consider many other parts of the system as well: elements of the natural realm – rainfall, storms, discharge of rivers, melting of snow; elements of engineering – roads and railways, bridges, dams, dykes; social structures – settlements and population in the flood area; administrative elements – flood control, communications and transport networks, medical help, organisation of rescue, insurance of property, and so forth.

---

* I do not here include geomorphology and climatology, as these branches of geography steer more or less consciously toward Earth Sciences, aside from efforts by some geographers of physical training to reconcile geomorphology and geography by developing the subject-matter of anthropic geomorphology [Nir, 1983].

This is no sub rosa return to determinism, but a call for the understanding that elimination of environmental challenges from a geographical study is a crucial decision which, in fact, decapitates the geographical approach. The dichotomy man/environment is only an *apparent* dichotomy; man and environment are two parts of the same system. The methodological difficulties of treating elements from both natural and social sciences together, as we will show later, can be overcome.

Knowledge of the physical environment is fundamental:

... actually, the human geographer stands in need of a knowledge of physical environment to precisely the same degree as does the archaeologist, the ethnographer and the economic historian' [Forde, 1939].

This is the minimum concensus required by the idea of geography as a unity of social and natural challenges. A standpoint ensuring the non-return to determinism can be found in Grossman [1977]: Geographers stress the theme of man's adaptation *of* Nature, whereas anthropologists investigate man's adaptation *to* Nature.

I wish to end my remarks on the theme of the *inseparability of man and environment* in geographical regional study with two examples. One is the existence of belts of hunger in the Sahel as we near the end of our much self-vaunted twentieth century. Are these belts of hunger caused by man in a difficult environment, or by the environment itself? The second example is the existence of a poverty region, Appalachia, in the heart of the U.S., where the common denominator to all parts of the area is a mountainous physical background. Neither example is an argument for determinism, but both require us to consider the natural environment as one of many elements making up a socio-natural system, as the backdrop to social structures.

The 1970s and 1980s will be remembered as years of hunger and food shortage in the semi-arid belt south of the Sahara [Mensching, 1986]; the blame has been laid to drought, on the one hand, and to social, economic, and [especially in Ethiopia] political disorder on the other hand. Blaming the environmental challenges proved unjustified, as there was no deficiency in rainfall over the period; it was soil overuse that was to blame. In an effort to enlarge the cultivated land area, even the nearly arid parts were ploughed. Heavy soil erosion resulted, and the fringe of the desert advanced because of this misuse of the land. *The fragility of soil* was the key element in the desertification; a faulty answer to an environmental challege was at the root of the problem. But in an economy of misery one does not expect a balanced response from an impoverished population. The 'drought' in Ethiopia was only a pretext used to mask the government's erroneous economic policies.

Appalachia is an area of rural poverty ranging across eight states, from New York in the northeast to Alabama in the south [Dilisio, 1983]. In these states the Appalachian areas have the highest rates of unemployment, illiteracy, deficiency in transport and so on. In the fifties the social structure of the region tended toward the abandonment of farmland, backward school systems, a population 'brain-drain', regional inertia, isolation. Despite an abundance of natural resources –

coal, natural gas, dense forests – the region contains the largest group of low-income population in the US.

The common denominator across the span of Appalachia is a rather hilly to mountainous relief which negatively influences transportation, agriculture and accessibility. When these *negative environmental challenges* were revealed by an integrated geographical regional study, it became the impetus for the launching of an important federal program to overcome these deficiencies by building a good transport network, implanting industry and improving education, with federal aid totalling some 1.2 billion dollars. This intervention slowed the rate of unemployment and negative tendencies of other demographic parameters.

This example demonstrates that natural challenges exist, as elements both positive [resources of coal] and negative [topography], and that they should be taken into account in planning, economy and management. The study of population in space without studying its environment ceases to be geography, becoming instead a geometry of distribution of settlements. The environmental constraints that confront people in a given locale cannot be ignored, even in the most technologically developed nation in the world.

The dichotomy space/place

'*A persistent error*'. The nature of geography is in the tension between place and space but, as I hope to elucidate, the dichotomy between these two concepts is only apparent, one of the 'persistent errors' or 'false dichotomies' [James, 1967]. The definition of 'place' and 'space' is the subject of vast disputation: space was considered by Aristotle a condition for the existence of things [James, 1967]; by Newton, as defined as objective reality but intrinsically void; by Descartes, as an essence of bodily substance; by Spinoza, as an attribute of substance; by Berkeley, as mental construct. Today, Claval [1984a] has defined 'social space' as 'area inhabited by a group, a product of human activity; it is the *abstract system of values and relations* that characterise a social structure.'

We have here, through history, a shift from a purely materialistic-physical definition – space as a part of the earth surface – to a sociological-spatial definition where man ['social group'] is the final criterion. At the base of the concept of space is the axiom of distribution: the earth's surface, being the arena of 'substances', 'things','social groups', becomes 'space'. On the other hand, points in this space, which have a discreet relationship with parts of the space, are 'places', which can be identified and characterised by *spatial coordinates*. It emerges therefore, that place and space are dialectically linked together: a place is a certain point in space, whereas space is composed of places and relationships between them, the basic one being called *distance*. Place, space and distance are the basic geographical elements which attend social groups and their activities; they are also at the origin of the diversity of geographical realities.

By definition, regional geography deals with places; but as places connected between themselves by relationships constitute space, *eo ipso* it deals with components of space. Following this concept, a discreet part of the surface of the earth, which we define as region, will be *both place and space*. Accordingly, the dichotomy between place and space is only apparent, as the definition of dichotomy is a logical division of a class into two opposite and mutually exclusive subclasses.

A logical interpretation of this principle in geographical practice is that both regional and topical geographies can deal with their subject matter in two directions: one studying the particular subject matter as a part of a whole, of space, by elaboration of general concepts, and the second dealing with a subject matter in its local aspects. Thus we can have a topical geography dealing with, say, geography of the iron industry as a socio-economic phenomenon, and elaborate general ideas about its structure and function; on the other hand, the iron industry of a particular state or district may be studied, in which case it is a part of regional geography if anchored in the regional system and its problems. In each case, the two approaches should be linked: the topical approach being based on local phenomena, and the regional approach relating to general phenomena. We can take as subject-matter of our study a systemic region, or, as coined by Whittlesey [1954], a 'total' region [cf. p. 111], dealing with its different elements and analysing the processes acting in it, i.e., the input, regional structure and the output of the region. Or we can enter upon the study of that region from the opposite direction, considering the acting processes as part of the spatial processes and learn from their adaptation to the local condition.

One may consider the relationship between topical and regional studies as a matrix [Reynaud, 1974]. There are two extreme positions – topical subject matter with a general approach, and regional subject matter with a regional approach – and two mixed positions – topical subject matter with a regional approach, and regional subject matter with a general approach. I see no difficulty in all of these being legitimate. Rather than insisting on dichotomy, *we can achieve through multiple approaches more unity in geography, precisely because of its diversity.*

As James [1967] explained it,

> ... all general concepts must be identified from the study of particular places; all study of particular places are made significant only through application of general principles. All topical studies must be done regionally, and all regional study must be done topically.

I do not insist on all studies being done in the same manner; I would certainly leave to the scholar some degree of freedom to apply his personal preference. But I do wish to reiterate that the dichotomy space/place is only an apparent one, and it is in fact a persistent error.

Places and space. All geographical activity orbits within the relations place/space: transport, migrations, production and supply, perception of distance, emotional perceptions such as 'fatherland', 'heimweh', longing for a beloved landscape. All of these are a direct or indirect product of the two-faced framework of our existence, place-space. As argued above, the one cannot exist without the second; but in an effort to eradicate the error of continuing to believe in an apparent, but erroneously perceived dichotomy, I shall devote yet a little more time to these concepts.

The focus of geographical activity [Jones, 1984] is space, which consists of places or locations and relationships between them. Location is but a point of orientation, a set of coordinates, which is absolute, but has no other than spatial meaning. It is a part of a geometrical – not a geographical – network. To become a part of a geographical network, a location must have attributed to it qualities, be they topographical, climatic, demographic, economic or other, by which a location becomes a place. According to this, we can distinguish two types of locations [Jones, 1984]:

Absolute location. As explained above, this is simply the geometrically defined point of location [Chapman, 1974]. It is unique, steady, and has a constant spatial relationship with other locations in space.

Relative location. This induces qualities existing on a certain location, which may differ from location to location. It is these qualities that are responsible for the diversity of locations and make of them *places*. They may be environmental, such as climate, geological structure, topography, and considered as being rather constant. On the other hand, some qualities of location are dynamic, changing over time: population, land use, settlement, economic value – in other words, the society which occupies a certain location and its activities. If the absolute location cannot change, the relative place can change through changes in its qualities: a village can become a town, an orchard a vineyard, a factory an industrial area. The geographical significance of a place is, therefore, in its relative location: '... *a place is a location with special social values*' [Jones, 1984]. If a location is unique in its absolute spatial setting, the *uniqueness of a place can be only relative*, according to the quantitative distribution of its qualities: size of population, degree of industralisation, education, architecture, religious groups, etc. But even the absolute location can be affected by relative considerations. For example, the absolute position of Great Britain did not alter after the discovery of the Americas, but its relative location did, from the outskirts of the 'Old' world to a central position between it and the 'New' world. What gives to an absolute location its geographical meaning and makes of it a place is the importance attributed to its environmental and social qualities. A place is a social function.

If we accept that places are functions of spatial differentiation, we arrive at the extreme notion that no two points on the Earth's surface are identical [James, 1967]. Of course, they cannot be identical from the point of view of spatial geometry, as each point is defined by different coordinates. But in the diversity of the earth's surface is a regularity which can be identified by geographical methods; this is the epistemological basis of *spatial analysis*, which studies places according to their common denominators, common relationships, common rules. The *regional analysis*, on the other hand, seeks the particulars of places. I see in both methods legitimate, if different, approaches to geographical research.

Spatial analysis – the study of the distribution of geographical processes and the laws governing them – is, as stated above, *only one method among several* for understanding geographical phenomena. I wish to deal with geo-graphy and not with geo-metry. Some decades after Vidal de la Blache stated – unfortunately, in my opinion – that 'geography deals with places, and not with populations living in them' [1913], Schaefer made the no less unfortunate statement that 'geography should pay attention to spatial analysis of the phenomena in an area, and not so much to the phenomena themselves' [1953]. In both cases, geography is reduced to looking at its subject-matter in space, at directions and arrangements. This is too narrow an epistemological basis for a discipline. Geography should be, first of all, 'a way of thinking' [Beaujeau Garnier et al., 1979], but *not only* a way of thinking. It should be mindful of its subject matter, which is the spatial distribution and diversification of geographical phenomena.

GEOGRAPHY – ART OR SCIENCE?

If geography is an art, then it is characterized by uniqueness, as is every work of art [Gottmann, 1957]. As art, it perhaps does not need theories and methodologies, only inspiration and original minds. This attitude can be traced to the French school of geography in the first quarter of this century, when the regional geographer's aim was to paint 'regional portraits'. Today, geographers no longer seek this goal; a portrait is static, whereas geographical interest is to explain as well as possible the dynamic aspects of geographical phenomena. The assertion that geography is an art was not followed by profound substantiation [Meinig, 1983]. Nevertheless, many geographers considered themselves artists, and many artists, especially writers, engage in geography [p. 72]. Geographical writing abounds in literature, and the study of the role of geography in literature, the analysis of its geographical elements – descriptions of landscapes, places, 'genres de vie' – is advancing [Meinig, 1983; Douglas, 1985]. Approaching geography as an art is not a call for geographers to become novelists, but a call for greater openness, for clearing away pedantic barriers, for toleration of geographical creativity wherever it may lead. Still, geography will deserve to be called an art only if and when a substantial number of geographer perform their work with artistry.

If geography is to progress toward science the question is, toward which science [Bailly and Racine, 1978]? Human sciences? Social sciences? Christensen [1982] thinks that it should be a human science; Claval [1984a] and Wirth [1984] among many others label geography 'social science'. Even if most geographers today claim that it is a social science – and this opinion is mirrored by the usual University practice of attaching the department of geography to the faculty of social sciences – some scholars still believe that '... geography is the ideal vehicle for the joining of hands of science and humanism, including the taking of moral positions on environmental and spatial issues' [Parsons, 1977]. Indeed, in some European countries, France for example, geography has been formally linked with history. Particularly in recent decades, historical geography has been gaining more and more ground, in countries whose histories span many centuries – France, England, Germany – and in relatively 'young' countries as well. Not only its subject-matter but also the trend toward using quantitative methods in research lead geography towards the social sciences [Amedeo and Golledge, 1980]. The other trend, toward humanism, is considered the 'dernier cri' in present-day geography; in fact, it is not new at all, although only recently has the geographical establishment conferred upon it the 'droit du citoyen'. Good geography has always been humanistic [cf. Pierre Gourou's 'Leçons de Géographie Humaine', 1975]. Perhaps these two interrelated trends in geographical orientation can be reconciled by the view that humanistic geography deals with *man as an individuum*, whereas social geography deals with *social groups and structures*. What seems clear is that determining the position of geography as science requires deeper insight into geographical epistemology.

It would be incorrect to say that in the past geography was always considered an art. Geography has included incontrovertible elements of measuring and precisely formalized knowledge, in what has been called 'mathematical geography'. The exact size of the earth and the measuring of the sun's apparent movements have been at the center of geographical observations since Thales of Miletos and Eratosthenes. Most of these mathematical elements became facets of cartography, cosmogeny, astrophysics, and cannot today be used as an argument for or against a scientific base for geography. This means that the fundamentals of geographical epistemology and methology must be elucidated, but before doing so some remarks on the nature on science itself are in order.

One can find several definitions of science as perceived by geographers. Ackerman [1962] gave a rather grand one: science is a creative activity of the human mind, which depends upon luck, insight, intuition, imagination, taste and faith, as do the pursuits of the poet, musician, essayist and philosopher. This condition of mind, which is preambulatory, *must be structured* by thought and experience in such a way that it reaches the creative stage *in accordance with the main epistemological trend of the discipline*. In geography, this so-called 'geographical thinking' means, according to Ackerman, *structuring the mind in terms of spatial distribution*. Another important element should form part of a scientist's mind: a highly developed sense of problem, *as science is essentially problem-solving*. Ackerman's resulting definition is *science is a quest for regularity underlying diverse events*.

Following is another geographer's essay to define the nature of science [Moss, 1977]:

Science is
a. *A body of knowledge*, regarded as very reliable, related either to a broad spectrum of subject-matter, or restricted to knowledge gained by a particular scientific method;
b. Particular activity engaged in by a *definitive group of people called scientists*;
c. *Particular method of achieving reliable knowledge*, methodologically and epistemologically.

Combining Ackerman's and Moss's definitions, I propose that *science is a quest for regularity underlying diverse events, executed by particular methods by a definitive group of people, unified by a certain epistemology*.

The current opposition between the positivist approach to interpreting and explaining facts and the theory-seeking orientation, or, in other words, the antagonism between the deductive and inductive methods of research, is as old as science itself. This was the rift between the Platonists and the Aristotelians [James, 1972]. The former felt that reality exists only in the mental image of phenomena and events; they were the *theory builders* for whom observed phenomena and observed events were but shadows of reality. The Aristotelians, however, believed it necessary to *observe* phenomena and *then* develop theory to account for what has been observed. Since the classical period there have always been both followers of the *theoretical-deductive* method and followers of the *empirical-inductive* methodology.

In geography, as in all living sciences, diverse orientations exist [Derruau, 1961]. But it seems to me that the opposing trends have only methodological significance. Isnard [1978b] attributes little importance to whether one begins research with facts and proceeds to theory, or verifies a theory by submitting it to a factual test. Science progresses through a dialectical process: *from observation to theory, and from theory to observation.*

If we examine the epistemological content of geography, we see that in most cases nowadays an adjective is attached to the word 'geography': historical geography, social geography, economic geography, regional geography, not to mention physical geography, bio-geography, geo-cartography [Barnes and Curry, 1983]. What is the unifying epistemology which makes of all these subject matters one discipline, geography? Would it not be more convenient to speak of geographical economy, geographical history, geographical sociology, to change the adjectives into nouns? If the 'geography' in these subject-matters consists mostly of *spatial distribution* – as many geographers indeed believe, seeing in geography a spatial science – then it is merely a branch of geometry, and not a science in and of itself [Reynaud, 1982].

I see geography as the science which explains the diversification of places called regions, comparing them with other regions and trying to find out rules and generalisations which can explain the processes within regions. The 'branches' of geography – historical, social, economic – are only those parts of geography which consider particular elements emerging from the study of regions and treat them according to their spatial distribution [Bartels, 1982]. Geography is not only the study of spatial distribution, but the study of diversification of space.

Though the growing trend is to divert geographical epistemology and geographical subject-matter, we should not see these tendencies as centrifugal if the community of geographers continues to share 'common ground' [Sauer, 1924] in *the will to understand better the spatial diversification of phenomena and the rules that form part of it.* What we consider essential in our study determines whether geography will figure as noun or adjective, as primary or ancillary. But '... we need a geography without adjectives' [Anuchin, 1973, 1977], a geography that can stand on its own.

In most disciplines unity lies in a common subject matter; in others, it is found in the unifying methods of study. Geography, according to Hartshorne [1958], belongs to the second group. The unifying method in geography is that of proceeding from the viewpoint of spatial variations. Let us consider this statement in more depth.

Today geography is fragmented into many branches, many of them new, many of them quite promising. But fragmented knowledge does not constitute a discipline or science, even if the research is in-depth and proceeds by a unifying method. Beyond these it is necessary to have a coordinating theory of analysis, of fact finding, of proofing of hypothesis, and so forth. A theory should provide a further understanding of the science as a whole. The logic of any science lies in the definition of its substance and in its methodological basis. I propose that the substance of geography is the study of environmental challenges and the behaviour of social groups facing them. If the old, nowadays unacceptable geography was interested in bio-,hydro- and lithospheres, and man was considered only as a species of the biosphere, my interest is to deal mainly with the fourth sphere, the *sociosphere* [Anuchin, 1973] *and its environment*, as a *socio-natural system*. All of our material resources stem from the environment, developed, of course, by human ingenuity, and it is unthinkable to study economic and social life without seeing in the two realms, human and natural, a world-wide system which develops through mutual interaction. The environment should be considered a medium for social development, not a deterministic framework of existence. The methodological base I propose, as anticipated earlier in this text, is the analysis of spatial diversification and the effort to discover rules affecting the distribution of spatial phenomena. The systems approach is a methodological as well as epistemological translation of this effort. The world-wide socio-natural system can be studied either by world-wide subsystems, be they in the physical, sociological or economico-political realms [topical geography], or by systemic regions [regional geography] – each method being the complement of the other.

At the beginning of this section it was queried whether geography is an art, as professed most, and the best, of the geographers of the first half of this century, or a science, as is the persuasion of most if not all of the present generation. The answer is to be found in practice. Most geography as presented today in publications has a scientific flavor, but there are differences in style, in use of metaphor. Some geographical descriptions of landscapes have clear artistic pretensions; in works on mental geography [Saarinen,1976] emotional values are important. If we accept the basic thinking and methodology proposed above – the systems approach – then science and art can exist together under one roof as two different expressions of geographical thought: one stemming from our consciousness, our ability to analyse and deal numerically with facts, the second stemming from our emotions, our inner attachment to things, not ever rationally explained but nevertheless existing and influencing geographical realities. Science and art are two sides of the human ability to abstract from daily life; as geography wishes to embrace as thoroughly as possible the human existence, any neglect of an important segment of human existence would certainly bias its potential. Methods of study can include both scientific and artistic ones, when we have a common epistemology and a common subject matter – the world-wide socio-natural system. What we need is steady rejuvenation: The process of discovery is the most concise manifestation of man's creative faculty [Koestler, 1959]. Our respect for the numeric aspects of reality should not overshadow our perception of non-quantitative moral or aesthetic values.

OBJECTIVITY AND SUBJECTIVITY IN GEOGRAPHICAL RESEARCH

The question, Is geography a science? leads to another question: does a substantive objectivity exist in science at all and in the social sciences in particular? Some hold this question to be entirely irrelevant: 'Objectivity is an illusion, but subjectivity is no more than another useless illusion' [Marchand, 1974]. Previous generations took as a given that a scientist should be 'objective', that is, that he should have no aim other than to come nearer to the truth, and no subjective biases or intention to achieve a goal beyond the objective treatment of the subject matter. Outsiders generally believed [Johnston, 1983] that science is an objective activity undertaken according to very strict rules. A certain set of values is attributed to academia: norms of originality, of communality in the exchange of information, of disinterestedness, universalism, and organised scepticism. Empirism assumes, that objects can be understood independently of the observer [Harvey,1969]. On the other hand, as geography deals with human beings, it seems rather inhuman to behave towards its themes objectively, as if they were lifeless objects.

Let us consider some definitions of objectivity and subjectivity. Wright [1966b] says that it is a misperception to see subjectivity as the antithesis to objectivity. Objectivity is a mental disposition to conceive of things; this disposition inheres partly in the will and partly in the ability to observe, remember and reason correctly. Subjectivity implies a mental disposition to conceive of things with reference to oneself, as they are affected by one's personal desires and interests. But if we accept Wright's definition of objectivity, it must be said that even his 'objective' observer has a certain personality which influences, volens-nolens, his activities and judgment. Unquestionably, human nature does affect science [Wright, 1966b]. Personal qualities such as originality, open-mindness, accuracy and scientific conscientiousness are not distributed equally. When present in balanced quantities in the individual scientist, these qualities can contribute to honest [I propose to use this word in place of 'objective'] research. But human beings *are* individua and even with the greatest effort to be honest, even if we use an objective mathematical language, our results will vary. In the exact sciences, for example, where no personal bias can be expected on ideological grounds, two scientists may approach the same subject matter from apparently the same common ground. But these two sciencists will have different temperaments, different training, different mental abilities, and it can be supposed that their studies will each bear the individual imprint of the scholar. Will this research be defined as objective despite this?

What all research requires is scrupulous adherence to procedure – not to bend logic, mathematics, statistics, or facts in any way, to have no aim inconsistent with pure research. But even in the exact sciences the choice of a certain variable depends on one's personal preferences, on previous knowledge – or on the backing of an institution which is ready to finance one's work. If this is true of the exact sciences, in the social sciences doubting objectivity is even more justified because their subject-matter is society, of which the scholar is a part. As such, he belongs to a certain social class, to a certain nation, religion, profession. Even if he is keenly persuaded that he has no intention other than to come nearer to the truth, his basic approach to society – which for him is a natural, non-biased socio-cultural milieu – has, nevertheless, a certain relationship to the subject-matter of his research. It is quite possible that the scholar is unaware of this basic, inherent bias. Another scholar, at the same level of honesty, but coming from a different social milieu, a priori has a different relationship to the same subject-matter.

Geographers are no exception. The love of one's homeland is a virtue common to all nations. Fostered from early childhood, it is one of the basic values of each individual. Quite naturally, for most geographers serving their nation through their work is a natural duty.

The love for one's land should not be considered as biasing the scholar's integrity but as an objective segment of a citizen's life. Subjectivity involves acting in the *knowledge that one is outside some general truth*. The scholar cannot judge his own objectivity or subjectivity. He considers his social approach to be objective, as he is part of a society where these approaches are part of a general concept, a *truth*, accepted as an objective value. We must distinguish, therefore, between an inherent, unconscious subjectivity [I call it the subjectivity of the scholar's social milieu], and conscious subjectivity where the scholar, intentionally, wilfully, follows a certain aim – whether a personal aim for success or a political aim to advance a certain thesis – and makes his work subservient to this aim. The first type may be called 'naive subjectivity' where the scholar is perhaps unaware that he cannot at one and the same time be a part of a society and outside of it, 'objective'. The second type may be termed 'intended subjectivity', where the scholar is both aware of his subjectivity and convinced that it serves his personal or political aims.

Examples abound in the different fields of study in historical, social or economic geography. Indeed, it is difficult to find any subject matter which is not the focus of conflict between nations or social groups. The historical geography of South Africa, for instance, is a history of conflicts between Dutch and Hottentots, Boers and Zulu,Boers and British, and, a fortiori, between the scholar members of these different ethnic groups. Social and economic geography also deal with relationships between different classes or groups of interests.

A scholar is a member of society and cannot pretend otherwise. If the scholar is aware of his inherent, or naive, subjectivity, he can overcome it through a special effort to include in his research arguments representing views different from his own.If a scholar proceeds with an intended subjectivity, he should explain this as part of his research; the reader will then know how to relate to the results of his work. If the scholar sees in his work a social or political mission, he must take the responsibility of presenting it as such, and not hide behind so-called 'scientific objectivity'. Scientific objectivity, in my view, can and should be applied only to methods of research and the scholar's intellectual integrity.

I would conclude by adopting Loewenthal's [1961] statement, that all knowledge is necessarily subjective as well as objective, as each image about the world is compounded of personal experience, learning, imagination, and memory.

USE AND MISUSE OF GEOGRAPHY

Misuse is the alter ego of use. Although the boundary between the two is sometimes clear, in many cases it is obscure, and the transition from use to misuse can be virtually indiscernible. These are among the most ambiguous terms in our language: one person's use may be seen by someone else as misuse. The notion is connected with iconographical and mental structures, and a single phenomenon – for example, military force, slavery, enrichment – may fall into both categories simultaneously, according to the observer's perspective. This also holds true in geography.

If I had to define the term 'misuse', I would say that it is a use – of an idea, method, product – other than as intended by its creator, discoverer or producer. In this sense, perhaps no ideas, methods or products exist that have not been or could not be misused. Even statements of truth, without being mistreated, can be misused, for example when accurate information is passed on to someone unauthorised to have it, whether in the personal, political, economic or military realms. Certain interests may consider not only disinformation, but information itself as misuse, even absent a misleading intention on the part of the informer. Geography, because of its diversity of subject-matter and some so-called 'self-evident' but mostly erroneous correlations, is perhaps more vulnerable to this happening than other sciences. Falsification of history is a rather common phenomenon, but a misuse of history can, in many cases, be laid bare: history deals with things past and because of this can be verified to a greater or lesser degree. Not only does geography, by contrast, deal with on-going processes, but the peculiar knowledge and expertise of the geographer influences decision- making which will influence future processes.

Geography – perhaps more than other social or natural sciences – has been misused throughout its existence. Geographers responsible for this, however, have not always been aware of misusing geography. On the contrary, they have often been persuaded that they are fulfilling a national, political or ethnic commitment. Their misuse has been appreciated as such only by their adversaries. What explains this peculiar position of geography? The likely answer, given by Wright [1966 b], is that geography involves the *study of territory*. Territory seems to be a basic concept of human existence:

Many of the largest and toughest roots of man's inhumanity to man are embedded in the circumstance that certain groups enjoy advantages over others because they occupy or control particular areas of the earth's surface. There are areal conflicts within every village, every state and every nation, and worst of all between nations. Neighbours quarrel over fence lines and wandering cattle; nations fight over boundaries and the control of vast territories.

The use of geography shuttles between the naive perception of facts and demagogic explanation of and argument over their meanings. As Wright [1966b] concludes, in spite of or because of this danger of geography being misused,

> ... those branches of science that deal with areas, their occupants, and those who control them in terms of their relative advantages and disadvantages can do much to lay bare the roots of human conflicts and the laying bare of roots is a necessary preliminary of their removal.

The ways in which geography is misued are manifold, including misuses of geographical knowledge, geographical concepts and geographical theories.

Geographical knowledge. Napoleon is believed to have said that the politics of a state are in its geography [Reynaud, 1974]: to rule, one must know the country's economy, physical infrastructure, demography, and other such data. Whether this is a use or misuse of knowledge will not be discussed here. Lacoste [1976] sees the geographical knowledge of a ruler as a means of exploiting the people: 'Géographie, c'est pour faire la guerre' [geography is for making the war] is the title of his book. He considers geographical knowledge as the power base for any political action; without geographical information, neither the ruler nor the ruled will be succesful. According to Lacoste, the ruler hides crucial information from the ruled and it is by their ignorance that the ruler's domination is possible. If the ruled were cognisant of the geographical situation, they would certainly react against the ruler. Thus an elimination of basic geographical information from the population is, in Lacoste's view, a misuse of geography.

Geography in the role of serving the country has been perceived as normal. Most geographical societies established in the second half of the 19th century were in the service of the colonial expansion of the particular nations [France, Britain, Belgium], and their function was considered a patriotic contribution of geography to the nation [Mackinder,1919]. Nowhere is geography so important as in the military. Every aspect of geography has military importance, but especially knowledge of the terrain. Topography has ever been one of the most common subjects taught in military schools. It is not surprising that most of the detailed topographical maps used in geographical teaching and research have been produced by military authorities. The detailed topographical map of France on the scale of 1 : 80.000 still bears the caption 'Map of the General Staff' [Carte d'Etat Majeur].

It is impossible to separate a military operation from its geography [Russell, Booth & Poole, 1954]. Military action affects all physical and biotic features, landforms, vegetation, transport, human geography, pattern of settlements. There is even a special subject in certain geographies, military geography, dealing with these subject-matters from the military standpoint. 'Never before [the second world war] had so much financial support been available for the development of new cartographical procedure' [Russell et al., 1954]. We leave to the reader's judgment whether this is a use or a misuse of geography.

Geographical conceptions. When geographical knowledge is misused by being hidden from someone to whom it is relevant, or transmitted to someone who is unauthorised to have it, the facts, at least, generally have not been falsified. Geographical concept, on the other hand, are misused both through inadvertent misinterpretation and intentional misrepresentation to achieve political, social or economic aims. Even the term 'applied geography' has diverse, politically engaged meanings [House, 1970]. It carries different connotations in different lands, across the range of political dogmas and ideologies. The motives for disinformation, or misuse of geographical conceptions, are mostly *group interests.* The truth may be wishfully distorted [Wright, 1966a] in order to mislead rival groups; the results of scientific research may be suppressed to prevent rivals from benefitting by them, or used to best rival groups. In one realm, at least, geography plays a role in the struggle for power. I refer here to *geography and politics.*

At the outset a distinction must be drawn between political geography and 'geopolitics'. Political geography is a legitimate field of geographical study. Elements such as boundaries, political regimes, and legislative proceedings [Gottmann, 1980] in their geographical significance are aspects of geography without which it would be impossible to understand socio-geographical processes, tendencies of development, and policies on the utilisation of national territories. The term 'geopolitics', however, takes us down the path toward the *enslavement of geography to political action*, using unbased 'geographical' conceptions as justification for political action. This attitude was widely used by nationalistic totalitarian regimes in the first half of the twentieth century; even today, the practice may be found in perhaps more subtle manifestations. The practice consists of developing tendentious theories on the periphery of reality, through a process of interpretation and dramatisation [Perroux, 1968a,b]. I will mention here a few of the many geopolitical theses.

People without space ['Volk ohne Raum']. If a people attains a demographic, economic and political power that overflows its national boundaries, this ideology justifies the conquest of territories of neighbouring nations which have not achieved a similar degree of development and are, according this ideology, 'empty'. This reasoning was the geopolitical justification of Japan's expansion in the thirties, when Japan invaded Korea, Manchuria and China. That this concept was unfounded has been amply demonstrated by Japan's history since the end of World War II, its population growing at a considerable rate and its ecomomy becoming one of the most dynamic in the world, all achieved *without* territorial expansion.

Small nation, encirclement [*'Einkreisung'*]. These are somewhat pathological perceptions of national existence [Perroux, 1968a,b]. Of course, smaller countries are subject to the danger of being swallowed by a mighty neighbour, but there are many examples – Sweden, Finland, Rumania, to name but a few – of lasting coexistence between small and great neighbours, the disparity in size of country simply one of many differences between them. For years, Cuba lived without this complex; now, all of its political behaviour is directed by it. The result is a 'Sparta complex', where all national resources, by apparent justification, are devoted to developing military power. In my opinion, the way to overcome fear of military conquest is the will, *on both sides of the boundary*, toward peaceful coexistence.

Ein Volk, ein Reich [*One people, one nation*]. This is a concept identifying political organisation with ethnic entity. This conception of nation-state, an accepted means of national emancipation, is a product of the nineteenth century and is considered the legitimate way to guarantee the freedom of national groups. Nazi Germany interpreted this idea as the right to annex the territories inhabited by peoples speaking German and considering themselves German. A tendentious misuse of this idea – a pretext, in fact, for territorial appetite – 'justified' annexation of not only German-speaking Austria but parts and later all of Czechoslovakia and Danzig; and, of course, it did not stop there. All this was predicated upon a 'geopolitical' ideology or 'science'.

Historical boundaries. This is one of the most dangerous myths in relations between nations, as if to the politico-social structures which existed a hundred, five hundred, a thousand years ago could be attributed actual meaning. If their traces are still recognizable by actual geographical facts, a common language, for example, then these boundaries can have a tangible meaning. But a purely historical event, of whatever importance in its time, cannot be interpreted now as still carrying the same political meaning. The fascist slogan of 'Mare Nostrum', meaning that all the territory that constituted the Roman Empire in the classical period was a legitimate 'Lebensraum' of fascist Italy, was one of the most eloquent misuses of an intentionally erroneous interpretation of boundaries. *Political boundary relates only to the political body which constitutes it.*

One would have expected the abandonment of the concept of 'geopolitics' now that history has given it the lie. But there seem to be are tendencies to use this framework of disinformation even today. Let me cite the content of the revue 'Géopolitique', no. 3, July 1983; none of the papers has the slighest connection with geography or political geography:

> Western Opinion on War and Peace'; 'Variations on the Theme of Peace'; 'Indian Ocean: Fragile Peace in a Dangerous World' [authored by a naval admiral]; 'What is at Stake in Poland'; 'An American in China'; 'Small Weapons and the act of War'; 'Towards the Strategic Defense of Tomorrow'.

In this issue, only one map – of the Indian Ocean – was reproduced, but there are many photographs of weapons, marching soldiers, the machinery of war.

THE SYSTEMS APPROACH AS A WAY OF THINKING IN REGIONAL GEOGRAPHY

Earlier the systems approach was mentioned as a possible means of formalising the geographical 'way of thinking'. This proposition is not at all new [Ackerman, 1953; Chapman, 1974; Leghausen, 1974]. But in practice the systems approach has not been used by regional geographers. As the aim of this text is to contribute to a revival of regional geography in such a way that it will challenge, and meet the challenges of, the twenty-first century, I propose the systems approach as both the epistemological and the methodological basis of regional geography.

The concept of holon

Two papers, one by Kimble [1951, written originally in 1946] and another by Schaefer [1953] spearheaded the criticism of and accusations against geography in general and regional geography in particular. Kimble's main argument was that it is not possible, either methodologically or epistemologically, to address in a single concept both socio-economic and physical elements, as they are absolutely different; a discipline consisting of both cannot exist. I have already answered this argument [pp. 7-9]. Schaefer's view – as the title of his paper indicates – is that geography, especially regional geography, deals with exceptional cases. Moreover, geography, according to Schaefer, is biased by '... unrealistic ambitions fostered by the unclear idea of a unique integrating science with a unique methodology'.

These critiques distantly echo the ideas of Durkheim, an opponent of Vidal de la Blache [Buttimer, 1971]. Durkheim sought to establish general sociological laws, whereas Vidal de la Blache to all appearances painted regional portraits. This simplistic distinction, however, does injustice to de la Blache, who considered the study of unique regions the basis for studies leading to general inferences [Buttimer, 1971].

First we must understand the terms 'unique' and 'uniqueness'. In geography, uniqueness may be either absolute or relative [Chapman, 1974; cf. also p. 12]. Absolute uniqueness in geography is the coordinates of a point in space; this is the base of spatial geometry. But, together with its absolute uniqueness, a point on the globe also possesses a relative uniqueness. This refers to the spectrum of qualities attached to the coordinate point, be they physical qualities such as altitude, climate, and soil, or qualities incorporated into it by a certain population's social, economic, cultural and political attitudes.

When I speak of the uniqueness of a certain region, I mean to its *relative uniqueness*; its absolute uniqueness is a fact which needs no further discussion. If I state that in the whole world there is only one Paris Basin, or only one Moravian Basin, I am referring to its relative uniqueness, relative to neighbouring and distant regions. The problem arises – and this was the influence of Schaefer's paper – when one interprets the term 'unique' to mean 'exceptional'. Let me emphasise that this is not always appropriate.

Bunge [1966] expressed it clearly: *uniqueness is a matter of grade*, and not an exceptional phenomenon to which physical or social laws cannot be applied. *The uniqueness of a region is in the unique organisation of interactions between the elements which exist in that region.* The objective elements may exist in other regions as well, but in different proportions, functions and interactions. In other words, *uniqueness is not synonymous with exceptionalism*. Of course, there are cases in which a certain element appears in such a unique quantity, that it becomes exceptional, and quantity becomes quality; but these are isolated cases and do not justify confusing uniqueness with exceptionalism. As labeled by Schaefer's paper, 'exceptionalism' led to the interpretation – fostered, certainly, by the opponents of regional geography – that regional geography deals with exceptional cases.

But the years of criticism against regional geography were also years of support for it [James, 1952; Robinson, 1953; Whittlesey, 1954; Gottmann, 1955; Hartshorne, 1959]. It was the 'quantitative revolution' [Burton, 1963] which brought to the opponents of regional geography, following Schaefer, the oportunity *to shift from place to space*. Unique phenomena, despite their uniqueness, can be treated as part of a larger whole and can produce lessons and conclusions of general value, if we understand that their uniqueness is relative – and if we have appropriate methods to treat them.

I can imagine no better example to illustrate the relative uniqueness of a geographical entity than the example of man. Each human being is a unique phenomenon on Earth, a phenomenon which cannot be duplicated and which has its own value and potential, whether in constructive or destructive directions. This view of man, we would hope, is common to mankind, even if there are those who regard human beings merely as 'gun-fodder' for their perverse fantasies. Man is an individuum, which means 'what cannot be divided' [Frankl, 1969].

This unique phenomenon, the human being, associates with other human beings to perform a common activities. Different individua join together to form families, clans, tribes, communities. They associate according to profession, habitation, and political, cultural, artistic or religious preferences. An individuum, in spite of and because of his unique qualities, becomes a part of a certain group or groups, which act as collective individua, and settlements, states and nations are formed.

At this point I wish to introduce the term 'holon', coined by Arthur Koestler [1969] as a unit which is, on the one hand, something final, complete ['holos'], but on the other hand a part of something larger. Of course, this term is of more ancient origin. It is

> ... the idea of a unified whole, where each of the parts has a definitive place, determined by the structure of the whole, and each part may be in turn such a whole in its own right. In 'Parmenides', Plato called it 'Holon' [Scolnicov, 1984, p. 214].

To illustrate the concept of holon, Koestler compares it to Janus, the Greek god with two faces looking in opposite directions. A holon is an entity on one hand introvert, directed within itself, a whole, a final entity; on the other hand, and at the same time, it is extrovert, outwardly directed, a part of a larger entity. Returning to our example, the human being is a 'holon'. Within himself, he is an individuum, an entity, unique in his physical and spiritual qualities, consisting of different elements which are realised in him as a whole; but towards the outside, the environment, society, the world, he is only one particle, one component of a large socio-biological entity. The relationship between these two facets of the personality of the human being, between his commitments to himself and his commitments to the wholeness of which he is a part, seems to be the central problem of human existence.

Region as a holon

Holon is an important instrument for understanding the concept of region. Indeed, I propose that a region should be considered a holon: inwardly, within the content of a certain structure, it constitutes a whole and something final for its components; outwardly, the region is one of the components constituting greater wholes, such as states and nations.

This approach of viewing *a region as a definitive entity and at the same time a part of a whole* is illustrated by the example of Greater Washington [Nir, 1987]. The existence of the metropolitan area of Washington, D.C., is evidenced daily by the activities of its population of more than two and a half million people – by their commuting from residential areas to the Washington downtown, by urban and suburban systems of transport, by economic activities and social relationship, by a continual building and expanding of the built-up area. The connections and interactions of its components are the base of its existence and definition. But, from an administrative point of view, Greater Washington comprises parts of Maryland, Virginia and the District of Columbia.

The relationships and interactions are facilitated by communications, by telephone communications among others. The term 'Washington Lata' or 'Greater Washington' [Maryland Suburban Telephone Register, 1984, p. 23] designates that area within which one may call every telephone number without needing to dial an area code, although different parts of this larger area are in Maryland [code 301], Virginia [code 703] and the District of Columbia [code 202]. When calling the same numbers from outside Washington Lata, however, their respective area codes must be prefixed. In telephoning from outside, Washington Lata does not exist: each number is either in Maryland, or in Virginia, or in D.C. [Fig. 1].

Thus, on the one hand, Washington Lata is a territorial entity, at least for the Telephone Company, where for local calls the boundaries between the parts of Maryland, Virginia and D.C. do not exist when you are situated *within* it. On the other hand, this entity is cancelled out, and each number again becomes part of a distinct administrative area, designated by a specific area code, when it is dialed from *outside* Washington Lata. This is exactly the meaning of holon: when viewed from the inside it is something closed, something final and defined, but when viewed from the outside appearing as part of something larger. Although this example deals only with telephone communications, it is clear that the phenomenon of Washington Lata is not limited to this aspect. The relationships and interactions in the region are intensive and manifold, and the boundaries of

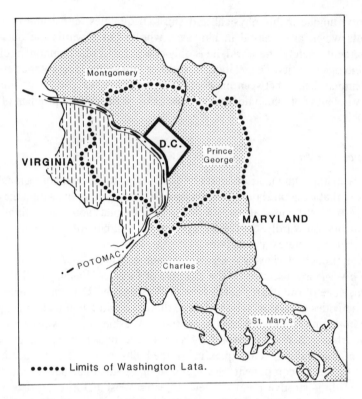

Fig. 1. 'Washington Lata' [Maryland Suburban Telephone Register 1984, p. 23].

it, as the telephone company uses them, are the inference and interpretation of them.

Just as man, a unique entity, can be agglomerated to different groups, the same is true of other unique entities, regions, which are composed of elements, measurable or appreciable. Therefore, the argument that the study of a region, being a unique phenomenon, cannot teach lessons applicable to other regions or other geographical individua, is invalid; but the way of study is not simplistic.

Region as a system

As a logical inference of what has been said until now, I propose to apply the systems approach as a method of research in regional geography. The term 'system' has different meanings, not always in harmony with that introduced by von Bertalanffy [Bertalanffy, 1951, 1962, 1969, 1971]. The word 'system' today conjures up an image of a battery of computers.

Despite the danger of such a misinterpretation of my intentions, I shall use the term 'system' because it expresses the essential: *an approach to phenomena not as to isolated items, but as components of a whole, where the relevance of the relationship between the components and their environment is anchored in the acting processes.* The difference between a sum of items and a system is that *the system is more than the sum of its components.* The overflow is not something quantitative, but the *relationship between the components.* Items lacking a relationship with the system are not part of the system [Bertalanffy, supra cit.; Bennet and Chorley, 1978; Chorley and Kennedy, 1971; Huggett, 1976, 1980; Ivanička, 1980; Miller, 1978; Wilson, 1980; Preobrazhenskiy, 1983].

One of the arguments against the systems approach in geography, as well as in some other social sciences, is that systems are built on the basis of numerical analysis and, therefore, phenomena that cannot be expressed by mathematical treatment cannot be included in systems. To overcome this difficulty, it has been proposed [Agnew, 1984; Huggett, 1981; Morgan, 1981] that the two types of systems be differentiated, hard systems from soft systems [p. 81]. A 'hard' system has clearly defined and rigid limits and structure, expressed mostly in mathematical language. But in geography there exist complex systems, the boundaries of which are not rigid, a system such as *drought*, for example. Within it are involved climatic, pedologic and agrotechnical aspects, as well as socio-economic components such as land use, nomadism, pastoral 'genre de vie', political decisions, financial aid, international cooperation, and so on. The boundaries of such a system cannot be rigid; by definition, such a system is 'soft'. A soft system cannot be expressed in numbers, but requires verbal description:

> Conventional techniques of system analysis are of limited applicability to societal systems, because such systems are, in general, much too complex and much too ill-defined to be amenable by quantitative analysis. It is suggested that applicability of system analysis may be enhanced through the use of the so-called linguistic approach, in which *words rather than numbers* serve as values of the variables [Zadeh, 1974].

Interestingly enough, in the same year that Schaefer published his critique of geography, Ackerman [1953] proposed a solution of the regional issue in the direction of the systems approach.

The systems approach can answer some of the basic problems of geography, and particularly of regional geography. It makes it possible to overcome the earlier discussed dichotomies in geography. Furthermore, by proposing an appropriate use of scale relative to the subject matter of a study [p. 83], this approach can overcome yet another dichotomy, the depth/width dichotomy, of regional research.

CONCEPTS OF REGIONAL THOUGHT IN GEOGRAPHY

By now of theories by the score
I was the proud possessor;
I hailed each new one as it came
And damned its predesessor.
And that is why at last I turn
Amid the paradigms legion
To study Hartshorne once again
And focus on the region.

Emrys Jones, *Area*, 1985, p. 169

In this chapter I wish to trace regional thought in geography. As with many other directions of inquiry and thought, the history of regional thought in geography cannot necessarily be likened to a trunk from which stem branches in new directions. Rather, ideas appear and fade and, like the tips of icebergs, only a small part of their total magnitude and potential is visible on the surface. Rarely is a new idea readily accepted and then further cultivated. In most cases, new and old ideas coexist for a certain time, perhaps the period of a generation. That one idea flourishes is not necessarily due to any transcending value it offers. That another vanishes may not be attributable to any particular weakness in its potential. Generally the social climate at any given time favours or disfavours certain ideas, and this can be the decisive factor in the longevity of any theory. Success is most assured, of course, when a favourable social environment hosts a personality which expresses the needs, aims, or expectations of that environment. At a certain favoured moments in history appeared a Humboldt, a Vidal, a Sauer, a Schaefer, a Gottmann, a Harvey who profoundly influenced the contemporary geographical thought. But even in periods which seemed overwhelmingly characterised by a distinct trend in geographical thought, there existed other, less pronounced, less published, less acclaimed trends: a theory from an earlier generation which persisted, a viewpoint of the rising generation timidly introduced. For the most part, the 'new' ideas have in fact been crypto-expressions of older ones; only a few can be appreciated as genuinely new ones. I do not intend here an extensive treatment of the development of geographical ideas, as this has been done elsewhere [Dickinson, 1970; James, 1972; Gregory, 1978; Johnston, 1983]. I intend only to outline briefly the history of thought in *regional geography*, which is, of course, related to general ideas of geography.

THE DAWN OF REGIONAL THOUGHT

Apparently a trend of gathering and simultaneous partition inheres in the structuring of human societies. As societies are formed by an aggregation of individuals into families, tribes, settlements, cities and states, a need develops to partition large social units into smaller, more manageable ones. Undoubtedly the first aim of regional partition was to administrate, to *reign* [from the Latin *regere*].

It is important to understand that at the dawn of 'Western' civilisation, geography – ergo regional geography – did not exist as a special science but as part of a general interest in acquiring knowledge. The Greeks may be credited as the first geographers, but it is certain that centuries before the classical Greek period the need to delimit areas was crucial to Egypt, where the annual floods of the Nile obliterated the established boundaries of fields [James, 1972]. Exact boundary lines were necessary not only to fix private rights of property but also to enable the central government to collect taxes. The practical purpose of 'regional' geography even then was to facilitate administration. This was, however, only one of the aspects of geographical thought in ancient Egypt. Some authors consider Egypt and Sumer to be the *cradles of geography*, as the places where measurements of *space and time* began: astronomical observations, the astronomical calendar, the development of geometry and mathematics.

Homer's *Odyssey* is viewed as being the most ancient text of regional geography, of a description of different parts of the world [James, supra cit.]. This is in no way a denigration of the literary worth of this work dating from the ninth century B.C., one of the earliest in human history. The *Odyssey* recounts the voyages of a hero returning from the Trojan War, travelling the fringes of the then known world of the Greeks. It contains descriptions of strange, sometimes fantastic landscapes, populations and creatures. After the passage of two and a half millenia, regional geography would still fulfil the same function: that of describing, purposing to explain, the new, apparently fantastic New World just discovered.

From Miletus, on the western coast of Asia Minor, came a major contribution to geography. This 'polis', a center of trade among the Greek islands, Asia and Egypt, was consequently an exchange point as well for masses of information on these countries which was both needed by merchants and obtainable from the traders who effectuated the transport of the goods. The two basic tendencies of geography – today labeled spatial and regional – were already present in the works of Miletian philosophers. Thales and Anaximander developed mathematical, physical and geometrical ideas on space and time, expresse in the sundial, the concept of meridian, perhaps a first map. By 475 B.C. *Hecateus* had collected oral reports and traditions brought to Miletus by traders and merchants. His aim was a description of the Earth, and he was perhaps the first [James, 1972] to make a regional division of the world into Asia and Europe, the boundary between them conceived as passing through the Hellespont, the Black Sea and the Caucasus. Apparently both conceptualisations of geography, the nomothetic one dealing with laws and general rules and the idiographic one dealing with information on places, are as ancient as geography itself.

A hundred years later *Herodotus* wrote his 'History', including in it descriptions of places he visited. He perceived geography as a means of providing physical background to the places where historical events occured. More than two thousand years after Herodotus this concept perseveres: in the French school, geography was connected with history, in a somewhat ancillary fashion, until the 1950s.

The greatest philosophers of ancient Greece, Plato and Aristotle, contributed mostly nomothetic ideas to geography; they did not deal with regional descriptions. Their ideas on the region-state [polis] were quite different [Gottman, 1984]. Plato's doctrine for the ideal polis was a territory

> large enough for the adequate maintenance of a certain number of men of modest ambitions and no larger; the population should be sufficient to defend themselves against wrongs from societies on their borders.

Plato's finest disciple, Aristotle, was the teacher of Alexander of Macedon.

> The Platonic model of the small, equal, self-sufficient and self-absorbed, juxtaposed territorial units may be easily opposed by the Alexandrine model of a vast expanding, pluralistic and cultural system, bound together and lubricated by the active exchanges and linkages of a network of large trading cities [Gottmann, 1984].

Plato and Aristotle differed in both their ways of reasoning and methods of research. Plato considered observations as mere copies of general, imperceptible *ideas*, the perfect predicates from which observations are deduced in a more or less *inexact* manner. Aristotle, on the other hand, reasoned that the best way to test a study is to confront it by *observation*. Sensory observations, however, could never provide an explanation; this was only possible through reasoning. Here lies the root of the dichotomy between scientific reasoning based on *deduction* and that based on *induction*. As proposed earlier [p. 15], this dichotomy should be considered only an apparent dichotomy. By a dialectical process of ideas and observations serving each our reasoning will achieve the better results. Indeed, the four essential principles of scientific explanation as formalised by Aristotele – nature of thing, substance of thing, processes by which they are formed, purpose of thing – can serve us even today.

Perhaps the most important regional geographer of the classical period was Strabo, who lived around the beginning of the Christian era and whose influence continued until the Enlightenment. Strabo wrote seventeen books, most of them dealing with geography. The purpose of his work was certainly to aid in the tasks of 'regere': 'it was intended to provide a text for the information of Roman administrators and military commanders, an administrative handbook' [James, 1972, p. 48]. This application of regional geography has remained unchanged to the present [p. 149]. As we near the end of the twentieth century, applied geography serves the regional planner, the development policies of governments an planning authorities. A major part of Strabo's work was devoted to detailed description of the various parts of the known world – eight volumes on Europe, six on Asia, one on Libya [Africa] – with two introductory volumes devoted to general considerations.

six on Asia, one on Libya [Africa] – with two introductory volumes devoted to general considerations.

The classical and medieval worlds were dominated by the geographical ideas of Ptolemy [second century A.D.], which survived virtually unchallenged for centuries. Books on geography did indeed appear, but these were generally compendia, or 'comptes rendus' of Arab, Jewish and Venetian travellers, rather than new ideas or concepts. Fresh developments had to await the age of discovery in the fourteenth and fifteenth centuries. There was one very important exception. The Roman geographer Marus Varro wrote a *cultural geography* in which he proposed, in our terms, a model of development of human societies through stages. It posited an initial civilisation of gatherers, transformed to pass through a pastoral stage, eventually reaching the agricultural and manufacturing stages. This 'model' was accepted as valid until the time of van Humboldt.

THE NEW WORLD

The sixteenth century was a century of tremendous, wide-ranging developments among the peoples of Western Europe. The discovery of the 'New World' changed with unprecedented vehemence the approaches to geography, transport, cosmogeny, mathematics, philosophy, religion – in a word, the perception of the Universe. Geography experienced a total revolution, as the ruling Ptolemaic concepts concerning the shape of the world could no longer be ascribed to. Everything was subject to a renewed scrutiny, as the explosion of information and experiences nourished new approaches. Novel facts, 'new' species of plants and animals confronted the theory of creation; Ptolemaic cosmography was measured against the Copernican-Galilean-Newtonian ones; religion reached unparalleled heights of fervency. The tasks of geography were many: elaboration of new instruments for navigation; preparation of maps; documentation of the new territories; and, of course, the writing of new 'cosmographies' including the newly discovered countries for the use of administrators, rulers, merchants and the general public. One of the most succesful of these cosmographies, copied and reproduced in many languages, was that written by Sebastian Munster in 1544.

The first work of geography in the modern sense was the *Geographia Generalis* of the German geographer Varenius, alias Bernard Varen. It was he who first established in his writings a relationship between the description of particular *places* and the description of universal *laws* that appear in all places. He denominated the former *special geography* and the latter *general geography*, what we today would call regional and topical geographies. He approached the concept of place as an interdependent part of a whole. Varenius was greatly advanced for his time [James, 1972, p. 126]. His book, written at the midway point of the seventeenth century, remained a standard in geographical education for more than a hundred years. He treated the physical aspects of a landscape as well as the physiognomy, art, culture, language, religion, cities, commerce, 'famous places' and 'famous men'. The chapters in his work are framed according to political units. Varen's ideas are very close to modern approaches in geography. His early death, at the age of twenty-eight, obviated his further development of these ideas.

The eighteenth century saw many 'cosmographies' or, as they were now called, 'universal geographies'; the word 'geographer' had replaced 'cosmographer' during the seventeenth century. Geography was not yet taught in universities but geographical texts were being written and propagated by Jesuit scholars [Dainville, 1940]. In these works, although the material was new, the method used was that of Munster, which was based, of course, on Strabo's tradition. Varen's concepts were not immediately followed, demonstrating that the propagation of new ideas is not necessarily a rapid process. Nevertheless some new ideas in geography came out of this period, the century which saw the development of natural history.

One of these new ideas was put forth in the 'Essai de Géographie Physique' by Philippe Buache, published in 1752. Its thesis was that the Earth's surface is composed of basins, delimited by mountainous systems. This very interesting idea was interpreted by a certain teacher, Johannes Gatherer, in a way which had a very peculiar influence on geographical-regional thought. Gatherer identified the drainage basin with natural region, in and of itself not a harmful idea. But he portrayed drainage basins, or 'natural regions', as *the organisational framework of geographical text*, including its human elements. This at base is an erroneous conception, as natural regions concern only natural phenomena. A region which includes in it both human and physical elements cannot be organised on the basis of physical elements alone. Political elements can, and in many cases do, override physical constraints. The delimiting of geographical regions by watersheds was a very easy procedure for drawing boundaries in unhabited areas, which were still abundant in that time, in the Americas, for example. But in succeeding generations, this notion of 'natural region', as applied to a system of landscapes and human societies, became one of the most dangerous and damaging ideas of modern regional geography. It is important to recognise that natural boundaries can limit only natural phenomena; regions, which are systems including socio-natural elements, can be delimited only by boundaries which take account of social phenomena.

In 1792 Anton Friedrich Busching published a work of six volumes, 'Neue Erdebeschreibungen' ['New Descriptions of the Earth']. Recall that each generation writes 'new' geographies. Even though his subject-matter division featured the Munster tradition of political units, Busching did introduce two innovations in this work: *population density* as a geographical element, and a prediction of the principle of *economic interactions* between countries on a global scale, some decades before the technology [steam engine, steam boat, etc.] existed for its realisation.

Two great spirits, Alexander von Humboldt and Carl Ritter, dominated geography from the end of the eighteenth through the first half of the nineteenth centuries. They heralded both the end of the previous age and the beginning of a new era in geography. The time had passed when a single personality could claim the ability to embrace geography in all of its branches, physical and human; the age of the 'renaissance' mind able to enter into physical as well as humanistic problems of geography was no more. This was the beginning of a new era, when the dimensions of the world became at once broader and narrower, broadened by new knowledge and information, narrowed by modern means of transport which lowered travel time and speeded communications. The gap, the dichotomy, between natural and human sciences now began to deepen. The world of the last quarter of the eighteenth century, when both Humboldt and Ritter were in the formative stages of their education, was quite different from that of the mid-nineteenth century as they neared the end of their works. Both men died in 1859, Humboldt at the age of 90 and Ritter at 80. These two giants were the last products of the humanistic tradition in geographical thinking as well as the first pioneers of modern scientific geography, based on observations, laboratory experiments, modern cartographical techniques and first-hand experience. They are rightfully considered [James, 1972] the *founders of modern geography*.

Humboldt and Ritter personify the two main trends in geography, one being the search for a world-wide basis for laws and rules, the other a search for a scientific explanation of the diversity of places. I do not wish to enter into an analysis of Humboldt's contributions, which were considerable both in natural history and geography. My aim here is to review the ideas of region and regional geography, and Ritter was the principal player in these areas of study. I do stress, however, that both although Humboldt and Ritter each specialised in one trend of geography, neither neglected the other: Humboldt produced regional studies, and Ritter wrote texts on general subject matter. Ritter labelled his work 'Erdkunde' ['geoscience'] as against Humboldt's 'Erdebeschreibung' ['geodescription']. Later generations call both approaches 'Geography'. Ritter employed the term geoscience to stress his scientific approach to his subject-matter. He would write a 'new scientific geography', not the traditional 'lifeless summary of facts, concerning countries and cities, mingled with all sorts of scientific incognities'. Strangely enough, Ritter's criticism of his predecessors is no different from that expressed a hundred years later concerning his successors.

On examination, Ritter's concepts of regional geography are revealed to be quite similar to those today considered the most advanced: regional geography is conceived of as *unity in diversity*; not an inventory, but an attempt to understand the *interconnections* and *interrelations*, that make the area a mutual ['zusammenhaengig'] association.

THE YEARS 1859–1953: REGIONAL GEOGRAPHY AT ITS PERIGEE

The year 1859 may be considered a landmark in geographical thought. It was in that year that both Ritter and Humboldt died and that Darwin's *Origin of Species* was published. From 1859 to 1953 geography was but a mirror of the prevailing ideas in other fields; it could, indeed, hardly remain indifferent to the march of scientific, economic, social, political and technological ideas and events. New methods of transportation and communication making the world 'smaller', the social philosophy of the working classes emerging amidst the industrial revolution, the trend toward ethnic groups forming nations, all were events reflected in geography and in its relation to the world and its problems. The influences of Auguste Comte and Charles Darwin, of Kropotkin and Reclus – even if the latter two found disfavour among the establishment – were in the background of the dominant developments in geography. To return to a metaphor employed earlier in this chapter, the trends in geographical thought may be compared to icebergs. A certain idea may dominate the horizon only because other ideas have been submerged, creating a deceptive surface impression. At other times, aspects previously submerged will pierce the surface and come into vogue. In other words, not one but several basic ideas coexisted, enjoying varying levels of popularity. During this period, regional geography was the predominant paradigm, but the seeds of other models can also be distinguished. These hundred years witnessed the shifting of geography from a more or less natural science to a rather pronounced position among the social sciences.

Social darwinism in geography

In the 1850s the physical environment was one of the basic issues in geography, the social sciences being only at the beginning of their development [Gregory, 1978]. This situation favored the theory of *development through time*, which is the basis of Darwin's theory on the origin of species. The theory of development through time was the most important external contribution to geography during the second half of the nineteenth century, along with the positivist way of thinking. Davis's theory on geomorphological cycles [Davis, 1899] was a direct outcome of these influences [Herbst, 1961], even to employing the terminology from the theory of development: young stage, adult stage, old stage. Darwin's influence in pedology, ecology and other fields is obvious, but his theory – or his followers' interpretation of it – also led to *social darwinism*, which became the dominant paradigm in geography from the middle of the nineteenth century through the 1920s.

Darwin's stress on the intimate relationship between organic life and environment gave impetus to organic interpretations of regions and states [Stoddart, 1966]. The idea of the connection an interrelationship between living beings and their environment later gave birth to ecology, leading to a paradigm of human ecology according to which geography is a *science of man and his environment* [Barrows, 1923].

The themes of selection and struggle were applied in a deterministic way to human and social geography. Determinism – the view that environmental influences are critical to human activity or habitation in a region – became the main force in explaining geographical phenomena [Huntington, 1924]. The idea of struggle and survival of the fittest found its way into geography. Many geographical theories stemmed from it, such as the concept of 'Lebensraum' ['realm of existence'],introduced by Ratzel [1882] and widely exploited by demagogues [p. 21]. Two generations passed before this paradigm of determinism was discredited by the Vidal de la Blache's paradigm of possibilism [Jeans, 1974; cf. p. 61]. Rather than according dominance to environmental influences, possibilism argues that not only can human beings adjust to the environment – a common phenomenon in the biosphere – but they are also capable of modifying the environment to their advantage, a lesson amply illustrated throughout history.

The idea of struggle and survival of the fittest was challenged in the last quarter of the nineteenth century by Peter Kropotkin, who considered *mutual aid* as the main force behind human survival; societies with mutual positive interrelations are the most successful [Kropotkin, 1902]. These ideas did not enjoy any particular acclaim among the academic establishment due to Kropotkin's political involvement as one of the leaders of the late nineteenth century anarchist movement. Nevertheless, his theory – together with the work of Elisee Reclus [Giblin, 1982] – significantly influenced some geographers and planners, such as Howard [1902] and Geddes [1915]. Social Darwinism persisted into the 1930s, even as the French school of possibilism gained ground. According to Herbst [1961], '... social darwinism emerges as the main cause of the sickness of American geography in the thirties.'

From physical to regional geography

Richthofen
Although his primary interest was geology and physical geography, von Richthofen should be given credit for many of the intellectual developments in regional geography. The purpose of geography, as he stated in his inaugural lecture at Leipzig on April 27, 1883, is to focus attention on the diverse *phenomena that occur in interrelation on the face of the Earth*, a concept not far removed from those holding sway among present day geographers. According to Richthofen, however, the way to understanding these phenomena is not particularly easy. To reach useful and reliable conclusions, a geographical study of any part of the face of the Earth must begin with a careful description of the physical features, and then move on the examination of the relationship of other phenomena to the basic physical framework. According to him, here the contemporary influences of the natural sciences and of the deterministic paradigm is evident. The highest goal of geography is the 'exploration of the relationship of man to the physical earth and to the biotic features that are also associated with the physical features' [James, 1972]. Throughout the century which followed, the stuy of this relationship became the basic geographical practice, until the shift towards the social sciences in the 1950s.

Richthofen's geography was a further development of the concepts of Varenius [p. 67] on *general* and *special* geography. There are, according to Richthofen, two steps in geographical research: *observation* and *explication*. Observations on which to build a framework of concepts must be made in particular areas where the features are unique; this is special or regional geography. But the knowledge achieved through observation must go beyond the description of unique features.

The next step is to seek regularities of occurrence and to formulate hypotheses that explain the observed characteristics. This makes the point that *good* classical geography was not satisfied with mere descriptions, nor was its goal the identification of unique features. It aimed higher, to reaching hypothess. Unfortunately, over the next hundred years the sheer proliferation of geographers and authors on geographical issues did not guarantee any particular progress toward the stage of explanation. Many geographies stopped at the first stage, the descriptive stage. The generalisation of this peripheral geographical usage seems to have generated the discredit of geography in general and of regional geography in particular.

The purpose of developing from observations of regional features general concepts concerning the global distribution of geographical phenomena – the Varenius *general geography* – was to understand the interrelationships between the diverse phenomena in the particular area. This branch of study was labelled *chorology* [choros = place], signifying that the interrelations between observed phenomena in an area should be examined on the basis of general concepts. To be able to see the world as a whole, it is necessary to examine smaller segments of the Earth's surface.

Hettner

From 1895 to 1937, Richthofen's concept of chorology was further developed by Alfred Hettner [Hettner, 1932, 1934/37]. His view was that *geography is a science unified on the basis of method rather than subject matter*. According to Hettner, the case of geography is parallel to that of history: the central issue of history is *the development of man on the earth in terms of time*, whereas the central issue of geography is *the development of man in terms of spatial variations* – again revealing the influence of Darwin's theory, the theme of the *development* of man. In his final concept of chorology, or 'science of places', Hettner does not accept the dichotomy between place and space. For him, the aim of chorology should be the understanding and knowledge of the character of regions through comprehending the interrelations among the different realms of reality, as well as comprehending the earth's surface as a whole [Stewig, 1979].

This immanent problem of geography – whether to be concerned by unique phenomena or to seek general concepts, whether to accept the idiographic or the nomothetic approach – occupied Hettner as it did contemporary historians and economists in their disciplines. Hettner insisted that geography is both idiographic and nomothetic, as all fields of learning should be. The passionate argument of recent decades over the idiographic or nomothetic nature of geography was already heating up a hundred years ago. Hettner's answer was clear: geography must deal with both unique and universal issues. But Hettner's views merely reflect the ideas of Varenius, expressed two centuries earlier: regional, Varenius' special, geography is concerned with the many unique characteristics of a place, yet this approach can be effectively pursued only when illuminated by the essential general concepts. As so aptly put by James [1972], since both Hettner and, later, Hartshorne [1951] made clear their view of geography as both idiographic and nomothetic,

... it is discouraging to find some writers [Schaefer, 1953; Harvey, 1969] who continue to accuse Hettner and his followers of defining geography as essentially idiographic.

Hettner's ideas influenced the 'regional model', which continued to dominate regional writing until the end of the 1950s [p. 90]. A study of the relationship between man and his environment began with location, examining geology, geomorphology, climate, vegetation, and natural resources, continued through forms of settlements, communications, distribution of population, and economy, and ended with political organisation. This rather rigid model was somewhat inconvenient; many authors reached only the first stage of investigation without continuing into the second, explanatory stage. Incomplete application of the model led to inventories or encyclopedic volumes, rather than to an analysis of spatial phenomena.

Vidal's concept of 'pays' and regional geography

Vidal de la Blache's concept of the region was a major development in the idea of region [Dumolard, 1980]. To understand it properly, one must take into account the period in which it was conceived. The activity of Vidal de la Blache in geography was the period from 1870 to 1919. At that time the population in France was rural, a people relatively stable and rooted to the land, and the French economy was for the most part agriculturally based. In this rather static situation, the object of geography was to describe and explain the *realities* of the landscapes, to show their pecularities and their spatial differences. To describe means to analyse the basic elements of physiognomy and to construct, by synthesis, the regional entity. This epistemological concept went hand in hand with the method-ological concept. As the aim was to identify different entities, with distinct 'per-sonalities', the basic methodology was mapping and the use of thematic maps. This approach, in its time, was the required answer to the reality of life, showing the historical development of segments of the country.

As had Richthofen, Hettner and other scholars, Vidal presented his basic ideas in his inaugural lecture of professorship at the Sorbonne, on February 2, 1899. His method of study was to examine the relationship between man and his immediate environment ['milieu'] by studying homogeneous areas, the *pays*, as the fruit of a long association between Man and Nature. Vidal's contribution of the historical approach was an important element in this concept. The result of this relationship is a generic region and a way of life, 'genre de vie'. According to Vidal, the 'genre de vie' is the basic factor in choosing from among the many *possibilities* nature offers to the inhabitants of the 'pays'. Hence Vidal's approach is *possibilistic, not deterministic* as had been the earlier approaches of German scholars [de la Blache, 1913].

The essence of Vidal's geography [Buttimer, 1971, 1978] was neither influence of nature *on man* nor the influence *of man* on nature, but the perennial tension between the external environment, the 'milieu externe' – meaning physically observable pattern and processes – and the internal environmemt, 'milieu interne' – values, habits, beliefs and ideas of a civilisation. The external provided a range of possibilities; the internal dictated the parameters of choice within this range. This is the Vidalian interpretation of possibilism.

Vidal's 'géographie humaine' studied three main issues [Buttimer, supra cit.]: distribution, density and movement of population; methods employed by man to develop his environment and his diverse civilisations; and transport and communications. His human geography was in fact a geography of civilisations. His regional studies were only a point of departure for a systematic treatment of 'genres de vie'.

Vidal's approach to geographical study stresses the *integrative* role of geography:

> ... the capacity not to break apart what nature has assembled, to understand the correspondence and correlation of things, whether in the setting of the whole surface of the earth, or in the regional setting where things are localised [James, 1972].

The influence of Vidal and his disciples and collegues, labelled 'la tradition vidalienne', influenced the geography of France and Europe for thirty years after Vidal's death in 1919. This school was characterized by a careful balance between the physical and human elements of geography, giving not the slightest hint of their constituting a 'dichotomy'. Physical geography was then and is still taught at the faculties of Humanities in French universities.

Even in the French school, however, the concept of regional geography slowly changed. From the 1930s, the region was studied not according to the original idea of a comprehensive investigation of all its elements – the model used by Ritter, Richthofen, Hettner, and others after them – but by focusing on a *central problem of it*. This problem might be the impact of the society on a landscape, type of land use [such as Pierre Gourou's specialised studies of the tropics], or a specific natural challenge such as a mountainous environment. Vidal's disciples and followers treated not only regional studies within the framework of 'pays' but also spatial aspects of the relationship between man and environment. They were perhaps forerunners of the environmental school of the 1960s. Subjects like 'Man and Mountain' or 'Man and the Forest' became quite characteristic of the French geographical school in the late thirties and afterwards. The greatest achievement of the Vidalian tradition was also its final one, the publication of the last 'Géographie Universelle', a compendium of the contemporary geography published between 1927 and 1948. This remains the 'magnum opus' of classical regional geography, studying the interrelationships between man and his environment on a regional scheme.

But it would be false and unjust to attribute to Vidal de la Blache the perception of the regional study as examining only the isolated, unique area. In his later works he attributed to the regional concept a wider role. He recognised the study of a certain region as being only a first step toward comparing it with other regions, which must lead to general lessons acquired from the study of particular regions.

He also perceived the growing importance of the city, as a focus of radiating influence on its surroundings:

> ... the formation of great agglomerations... constitutes... the most effective machine which humanity has thus far succeded in contriving. Their serried cohorts can be drawn upon... by an area encircling them like a halo [de la Blache, 1926, p. 158].

We must remember that in Vidal de la Blache's lifetime the galoping urbanism that characterised the distribution of the world's population during the second half of the twentieth century was only in its early steps, other than in parts of the US and Britain perhaps. Nevertheless, the Maître had understood its importance. Fifty years after Vidal's death the study of the urban environment, including its social and economic processes, would become the most dynamic aspect of geography.

The two apparently opposing trends in geography, the regional and the topical, were heavily discussed at the turn of the century. In opposition to Vidal de la Blache was the teaching of Durkheim [Andrews, 1984]; as Vidal de la Blache was one of the founding fathers of French geography, Durkheim was one of the founding fathers of French sociology. They shared a common field of studies [Berdoulay, 1978], the reciprocal relations between society and environment, but their methods of explanation were quite different. Durkheim stressed the need for systematic studies of society leading to general conclusions and laws, and insisted on rigourous concepts, on a unitary view of social sciences. Vidal suggested, as stated earlier, two complementary modes of analysis: the regional one studying the intricate connection between inner and outer 'milieux' in certain places, and a systematic one focusing on specific elements of 'civilising' the environment. Opponents of Vidal's geography claimed that it is limited to the regional, to the unique, and that the study of a region is the proper goal of geography; they also adopted Durkheim's approach of general laws and conclusions [Schaefer, 1953]. But as early as 1896, and again in 1898, Vidal de la Blache clearly expressed his view of the need for generalisation, and that the region is only a part of the geographical study:

> Intellectual curiosity drives us to relate the isolated detail, in itself unexplicable, to a greater whole which sheds light on it. When they draw their inspiration from the higher level of generalities, local studies gain a meaning and scope which greatly exceed the particular case under consideration [An. de Geogr., 1896, pp. 141-142];

> Geography must become a science which analyses, classifies and compares [An. de Geogr., 1898, p. 97.; both translations by H.F. Andrews, 1984, p. 327].

The geography of Vidal de la Blache, even when dealing with 'unique' phenomena, did not consider the unique phenomenon, the region, as an ultimate goal but as a *part of a greater whole*. Generalization was an essential part of Vidal's concept of geography.

Landscape as explained by Carl Sauer

In the last quarter of the nineteenth century the idea of 'Landschaftskunde', the science of landscape, emerged in reaction to the chorology, or science of places, of Hettner. Its protagonist was Otto Schlueter [Dickinson, 1969]. His approach to regional study was that of historical geography. Before the coming of Man, there existed a 'natural landscape' [Urlandschaft]. This landscape was transformed into Kulturlandschaft, a landscape created by human culture or civilisation. The major task of geography, according to Schlueter, is to trace the changes thus wrought.

In the between-the-war years of 1919 to 1939, when the main trend in regional geography in France and other parts of Europe was formalized by studies of 'pays', the 'Landscape School' of Carl Sauer, influenced by German geography, developed in the US. It was Sauer, in fact, who translated the term 'landschaft' into 'landscape'. He should be credited with many modern concepts anticipating the 'post-quantitative revolution' thinkers. One example is the concept of basing geography on phenomenological philosophy:

> All science may be regarded as phenomenology... Every field of knowledge is characterised by its declared preoccupation with a certain group of phenomena [Sauer, 1924].

Each science, according to Sauer, is in its first stage *naive*, in so far as it takes as given that part of reality that is its field of study. In its next stage it becomes *critical*, undertaking to determine the ordering of and connections between phenomena. The landscape, or the area, is the subject matter of geography, being a naively given section of reality, but the phenomena that make up an area are not simply asserted but are *associated or interdependent*. To '*discover the connection of phenomena*' is the only scientific task to which geography should devote its energy.

The sad truth is that many regional geographers remained in the naive stage of their investigations. Such geography cannot be intellectualy satisfying. It was not the geographical concept that failed, but its application. Had the generation of the fifties read more attentively Hettner's, Vidal's and Sauer's work, perhaps the divorce of regional from topical concepts would not have occurred: many of the 'modern' concepts were already apparent in those of Sauer.

To Sauer, landscape is an area made up of distinct associations of forms, both physical and cultural. But a landscape is not an isolated phenomenon: '... its identity is based on constitution, limits and generic relation to other landscapes, which constitute a *general system*.' I wish to heavily underscore the last two words. Sauer published this idea in 1924 [Leighly, 1963], years before the systems approach became popular and accepted as a way of thinking in the sciences. Classical geography applied the system approach unconsciously, naively, long before other sciences, as concretely expressed by Sauer. Perhaps because of the 'embarras de richesse' of the phenomena existing on the world's surface, geography was overly occupied with the first, descriptive stage of its scientific mission.

Sauer's concept of a system is basic: 'One has not fully understood the nature of an area until one has learned to see it as an organic unity'. Organ presupposes function. A landscape, or area, or region, must create processes in order to fulfil its function. Sauer sees in the description of a landscape the revelation of its genesis, the processes which created it and are still active in it:

> ... beginning with infinitive diversity, salient and related features are selected in order to establish the character of the landscape and to place it into a system.

This sentence requires careful attention. It heralds the current concepts of subjectivity in geographical research. 'Related features are selected': related *to what*? selected *by whom*?

Sauer does not see the purpose of isolated, unrelated descriptions. As we shall see, he considers the unique, the special, relevant only when it figures as a part of a system:

> No valley, no city is quite like any other city or valley; in so far as these qualities remain completely unrelated, they are beyond the systematic treatment, beyond the organised knowledge that we call science. No science can rest at the level of mere perception. There is no idiographic science, i.e. one that describes the individual merely as such. A definition of a landscape as singular, unorganised or unrelated, has no scientific value.

The system approach, seeing a given landscape, area or region as a subsystem of a larger system, is the methodological and epistemological lever which can move regional or landscape geography into a scientific path. But before this new approach could be fully developed, regional geography had to traverse its 'via dolorosa' of the third quarter of the twentieth century through the 'quantitative revolution'.

THE 'QUANTITATIVE REVOLUTION': REGIONAL GEOGRAPHY AT ITS APOGEE

> *Even the greatest ideas of science are nothing more than working hypotheses, useful for purposes of special research but completely inapplicable to the conduct of our lives or the interpretation of the world.*

E.F.Schumacher, 1975, p.87

The period after World War II was marked by profound changes in all realms of human activity: technology, transportation, economics, social studies, humanities and science. Today, forty years after the beginning of this trend, a new, 'post-industrial', 'informational' culture is becoming the common denominator to important parts of the world. A significant number of geographers enthusiastically, zealously even, followed the general trend in the social sciences of quantification. In geography the years 1953 to 1963 have been labelled [Burton, 1963] the 'quantitative revolution', but there were harbingers of quantification even earlier [Renner, 1935; Unstead, 1935]. The 'revolutionaries' rapidly became the ruling class, and the quantitative approach was accepted by the geographical establishment as the mainstream of the discipline. In some ways, geography in the 1960s was characterized by a *regime of great numbers*. This is not to say that there were no important scholars working in the qualitative direction. During that decade, however, their voices were not often heard, their works were not readily published or sometimes not published at all, and their research did not benefit from the enormous grants which the 'regime of great numbers' awarded its protagonists.

But '... the changing currents of geographical thought are not always manifestation of progress' [Mikesell, 1976]. Some ideas persist, others are embraced for a while and than abandoned. By no means all the ideas that are abandoned have been disproved by objective standards [Mikesell, 1976]; they may simply be discarded because they are no longer fashionable. Even in the period of the quantitative revolution, the regime of the great numbers was not total, despite its totalitarian attitudes; some of the basic ideas of geography other than those quantitatively measurable, including some important works in regional geography, were produced in that period.

No theory, even one figuring in the quantitative revolution, is born of spontaneous self-generation [Isnard, 1978]. It can only result from a lengthy germination and maturation, from reflection and a basic working out of factual material. The principal innovation of the 'quantitative revolution' was not the use of measurements, computation, and statistical treatment of geographical data; these were only its tools. Its main intended epistemologic contribution was a move away from the study of a particular place to the study of processes of spatial dimensions, from a static description of a region to a dynamic explanation of the society in this region. The sixties generation was dissatisfied, however, not only with the object of study. They also quarreled with its philosophical background, the empiristic philosophy of the regional paradigm [Johnston, 1983] which argues that what we experience is what exists, and that the role of methodology is to present the facts as experienced. By contrast, the topical paradigm was based on the positivist approach: that the knowledge gained, admittedly, through experience should enjoy general acceptance as credible evidence only after the factual statements are verified by scientific method.

As discussed in the previous chapter, the 'new' ideas, by and large, were not in fact new. They merely emerged from a state of latency, becoming the current vogue in geographical thinking. What actually happened is that the trend toward systematic research in the tradition of Durkheim [p. 82], invoked by Schaefer in 1953, coincided with the coming of computers which made possible the coordinated treatment of large numbers of variables. At the root of the 'quantitative revolution' lies the dissatisfaction with idiographic, i.e., regional, geography, but as explained elsewhere, this interpretation of regional geography was erroneous.

According to Burton [1963], the 'quantitative revolution' resulted from the impact upon geography of the work of non-geographers,'... a process shared by many disciplines, where an established order has been overthrown by a rapid conversion to mathematical approach.' The traditional French school of regional geography was preoccupied with empirical field work and '... had little appetite for methodological controversies' [Ley & Samuels, 1978b, p. 11]. Sauer and his students, while comprising the American school closest to the humanist content of French regional geography, also had little interest in epistemological controversies. One of the failures of the humanistic-regional position was not taking an epistemological stance against positivism in the sixties, as personified by the quantitative approach. The impression in the early sixties was that regional geography was being replaced by the positivist, quantitative methods. In fact, it was humanistic geography, oriented to the unique, to the individuum, which was under siege.

POSITIVISTIC APPROACHES AND TOPICAL STUDIES

Schaefer's paper [1953] is considered the manifest of the quantitative revolution. Ten years later, Burton [1963] stated that 'the quantitative revolution is over'. Within ten years, therefore, a major change took place in geographical thought and practice. The consequences as to both its protagonists and its opponents and critics are still discernible in the leading forces of the current geographical thought.

Schaefer's thesis was discussed previously [p. 23] with a view to demonstrating that his treatment of with the ideographical and 'exceptional' concept of regional geography was inaccurate and did injustice to the classical concept of regional geography. Nevertheless, Schaefer's thesis became the leading concept in discrediting regional geography in the eyes of its followers. The main issue became the *replacement of place by space* as the central axis of geographical practice. Embracing the computer, this approach brought about a quantification of geography and widespread use of statistics and models, as the shift from place to space occured in the same moment that quantitative methods borrowed from natural and social sciences were introduced into geography. From an epistemological point of view, the quantitative revolution may be viewed as the combining of positivist approaches with topical [systematic] subject-matter, and from the methodological point of view, as a method of quantification and computerisation.

Positivist approaches

Positivist approaches are basic to the 'scientific method'. They involve making empirical generalisations, with statements having the force of universal laws [Johnston, 1983]. The origin of positivism may be attributed to Auguste Comte [1829, 'Cours de philosophie positive'], who defined five postulates for a scientific method [Gregory, 1978]:

1. *Reality*: knowledge should be guaranteed by direct experience of immediate reality; *there are no abstract forces outside the direct experience*;
2. *Certainty*: unity of scientific method, which gurantees accessibility to all scientists and ensures the *replicabality of their observations*;
3. *Precision*: scientific method is the construction of *theories*, whose results *can be tested*;
4. *Utility*: all scientific knowledge is utilisable;
5. *Relativity*: scientific knowledge is unfinished and relative; scientific progress is paralleled by social progress.

The 'Vienna Circle' of the 1930s took major strides in the development of positivism. In the late thirties some of its protagonists went to the US and influenced the positivist school in that country [Johnston, 1983]. The development of positivism led to the formalization of *logical positivism*: verification by a logical processing into elementary statements and by the crucial experiential test of comparing the hypothesis with reality. Logical positivism includes three other doctrines:

a. *scientism*: the positivist method is *the only true method* of obtaining knowledge; other methods lack relevance because they cannot be verified;
b. *scientific reliability*: logical positivism provides the method for finding rational *solutions to all problems*;
c. *value-freedom*: scientific judgements are objective, independent of moral or political constraints.

I shall examine these doctrines later as well as criticisms of them. These principles, entwined with a mathematical and statistical apparatus, formed the basis for the quantitative revolution in geography. Consequently, certain subject matters were favored and accepted, others disfavored and excluded from geographical research.

Problems and questions which could not be quantified – beliefs, initiatives, ideologies – could not be the subject of a positivist treatment.

The second springboard of the quantitative revolution epistemology, besides the positivist approach, was its total devotion to topical [systematic] studies.* Throughout the history of geography topical studies have existed alongside regional ones, only bearing different labels, for instance the 'special' and 'general' geographies of Varenius [p. 32]. Studies on topical themes, such as human groups, economic activities such as agriculture and mining, environmental issues such as life in the mountains, nomadism, types of settlements, and so on, have always been an important part of geography. The distinctive mark of the generation of the 'Quantitative Revolution' was *the refusal and negation of the right to exist* of the regional approach, in favour of topical themes. Now the only acceptable treatment of a geographical topic was not on a regional but on a national or global scale, with the purpose that of learning the processes behind the facts. Perhaps this shift was impelled in part by the emerging modern-day society: the rise of the city, accompanied by the various processes of mass transport, mass production, communications, industrialisation, demographic and social changes of the society. The need in this new world was not to describe the development of a region through history, but to deal with acute social, economic and planning problems.

Although Schaefer's paper is accepted originating the quantitative and topical revolution in geography, dissatisfaction with the products of regional geography had been felt before. Ackerman [1945] viewed regional geography as undermining systematic studies, because geographers were amateurs in the subjects on which they published. Kimble [1951] considered natural and human processes a dichotomy whose treatment within a single discipline was unacceptable [p. 7]. On the whole, the attitude of the positivist-topical school was intolerant and did not admit of a pluralistic approach or coexistence with other theories. Studies in regional geography were disfavored in publications, in congresses, in topics for doctoral theses. This approach to regional geography lacked justification, for during the years of the quantitative revolution there occurred simultaneously a clarification on the standpoint of contemporary regional geography. Hartshorne [1959] clearly defined the task of regional geography as that of observing phenomena, classifying these observed phenomena in terms of generic concepts, obtaining maximum understanding of their scientific interrelationship, and arranging them into orderly regional systems. Between this paradigm and the 'exceptionalism' and 'uniqueness' so superficially charged by Schaefer and his followers exists a great gap or, perhaps a singular lack of sincerity. Most among the new generation, perhaps because of the statistical education which they enjoyed, opted for the quantitative methods of models and law making. This was especially true in the US and Britain and in the countries within their sphere of cultural influence.

---

* To avoid confusion, I propose for this branch of geography the term 'topical', as the term 'systematic' brings to the mind the noun 'system' and associations with it.

THEORIES CONTRIBUTED BY THE QUANTITATIVE REVOLUTION

Because they championed a new trend in geography denying the traditional regional paradigm, the proponents of the quantitative revolution had to define their realm of study. The new identity needed defending in the form of promoting a central field of study. The aim of the positivist-topical approach was defined by Harvey [1969] as '*a study of distribution of objects and events in space*'. So geography became 'spatial science' or 'spatial social science'. A detailed study of spatial science lies beyond the purview of this text, and I will cite only briefly the principal paradigms and theories which emerged from the quantitave revolution.

Central place theory. This theory, in fact, was conceived decades before the quantitative revolution. It found popularity within the Anglo-Saxon world, though, only after the works of its proponents were translated from the German [Christaller, 1966; Loesch, 1954].

Land-use theory of von Thuenen. This theory of hierarchies among cities and towns was of an even earlier, nineteenth century, origin [Von Thuenen, 1966], but came under consideration by geographers only during the quantitave trend of the 1950s.

Two themes were the genuine products of the quantitative revolution. One was the concept of *city as social area*. This subject matter arose both as a consequence of both rapid urban growth and the increasing interest among the American public during the late fifties [Johnston, 1983]. The second was *spatial interaction*. This was the main, truly original contribution of the quantitative revolution to geographical research.

As will be seen, the trend towards quantification led, perhaps inevitable, to a schism within and finally a breaking away from geography, because of the rule of *dealing only with measurable subject matters*. Social geographers' growing interest in economics brought about the development of regional sciences, which became a distinct discipline, independent of geography, although anchored for administrative convenience in the departments of geography at some universities [p. 55].

APPRECIATION AND CRITICISM OF THE 'QUANTITATIVE REVOLUTION'

After fifteen years of progress and conquest, the quantitative approach began to lose steam [Marchand, 1974]. I would present here an appreciation of the quantitative revolution in geography, and its influences on the 'post-revolutionary' generation. Its epistemological contributions were presented above, albeit in a very summary fashion. From the methodological point of view, it impact was that a high degree of mathematisation penetrated geographical research in all fields [Kohn, 1970], and as a result geographical argumentation became more precise. A further contribution was the use of models and model building. One of the basic elements of the quantitative approach was probabilism – replacing the determinism and possibilism of earlier generations – based on statistical laws, whereby human actions are interpreted as *rational decisions of 'homo economicus'*. The prototype of homo economicus is a human being deciding on his course of action purely on the basis of economic and rational criteria. This concept of man was the first to be attacked later by opponents of positivist thought [Haegerstrand,1973]. The positivist approach of the quantitative revolution – i.e., the belief that mathematical expression and quantitative analysis in and of themselves guarantee objective treatment of real facts – was questioned even at its perigee [Burton, 1963]:

> geographical studies are not descriptions of the real world, but rather *perceptions* passed through the double filter of the *author's mind* and his available *tools* of arguments and representation.'

An intellectual revolution is the first step toward counterrevolution. The revolution ends when its ideas are either overthrown or, if accepted, at the moment that they are modified by new ideas and what was formerly revolutionary becomes a part of the conventional wisdom.

In 1962 Rachel Carson published *The Silent Spring*. The publication of this book marked a new point of departure in man's relations with his environment. It was the conscious beginning of 'environmentalism', of seeing the environment as a part of human activity – and finding that activity to have largely harmful effects. The enormous changes wrought by man in the twentieth century – damming of rivers, strip mining, extensive ploughing of 'virgin soils', building massive road networks, polluting water with chemical residue and oil, destruction of flora and fauna – were suddenly perceived not as an aspect of the triumphal march of human development, but rather as a deplorable waste of natural heritage. This world-wide movement could have become a cardinal issue for geography, which for decades had claimed that its central problem and its epistemology were anchored in the relationship between man and his environment. This concept could have been a centripetal force to strengthen geography after the apparent failure of the regional concept. If there existed a subject-matter fitting the classical geographical paradigm, it was this integrative problem of conserving the environment while preserving the supply of raw materials and food. But at the very moment when interest in the environment peaked, geographers were occupied with the discussion between quantitative and qualitative approaches, between systematic and regional trends of study, and the subject-matter of Man and Environment by-passed geography into the newly emerging, interdisciplinary 'Environmental Sciences' [H.Sternberg, pers. communication, January 1985]. One of the greatest deceptions in the history of the quantitative revolution occurred at that crucial moment, when all effort was devoted to victory over 'traditional' geography instead of joining the traditional with the new epistemological context, in the most favorable of circumstances in terms of social interest.

Another point of criticism against the positivist approach, using the methods of 'great numbers' for statistical treatment, is that from the need to create averages of the variables used in statistical analysis, the variables are reduced to use only in relation to the common quality needed in the model, and then are averaged and incorporated in a statistical model. The qualities not used in the model, which may be of cardinal importancee for the discrete variable, are disregarded as 'noise'. But it is precisely this noise which is the individuality of the variable [Nir, 1985]. This 'dehumanisation' of variables was one of the central points of criticism of the quantitative-positivist approach and resulted in the crisis of 'humanistic' geography [p. 52].

Regional geography during the quantitative revolution

The years 1953–1970 may have witnessed the shift from place to space but, at least during the first decade of this period, regional geography was also very much in evidence. Some of the finest works in regional geography were published then, Gottmann's 'Virginia at mid-century' [1955], for example, as well as some of the most important theoretical works on regional geography in the US, such as Whittlesey [1954] and Hartshorne [1959]. The opinions discrediting the nature of regional geography [e.g., Kimble, 1957b; Schaeffer, 1953] were counterbalanced by positive assessments [Robinson, l953; Hartshorne, 1955]. The most comprehensive treatment of the regional approach ever made in the US, and perhaps in the world, was that of Hartshorne, developed over the years [Hartshorne 1939, 1955, 1959]. But the modern opinions on regional geography were neglected by the young generation of freshmen, who were fascinated by trends in other social sciences [Gould, 1964].

Ackerman [1953] brought fresh ideas to the regional concept, in accordance with the new conception of geography as social science. He defines region as a

composite phenomenon which has resulted from man's occupance of the Earth. The focus of regional geography should be a definitive description of the social and physical forces that shape a region in time, as they react each on each other.

Ackerman's paper contains indications of the future approach to region as a system. He speaks of 'hierarchies of regions'; hierarchies anticipate a perception of systems and subsystems. He sees a region as built up from components; it is 'not the totality of a region [that] is a practical goal for analysis,but the *key features* of it'. It is quite interesting to read a paper published in 1953, the very year of the beginning of the Quantitative Revolution, making the statement:

No geographer in these days could deny the existence of a region; there are some who despair of ever reaching of full understanding of a region.

Perhaps the most questionable difference between the protagonists of the quantitative-positivist approach and 'traditional' geographers was the pluralism of the latter and the uncompromising approach of the former. The positivists admitted of no 'droit du citoyen' of the regional approach, whereas the 'traditional' geographers advocated the right to exist of both the regional and topical approaches [Whittlesey, 1954]:

Geography may be considered a monistic discipline studied by two approaches: study of topical fields involves the identification of *areas of homogeneity* – the regional approach; study of regions that are homogeneous in trends of special criteria, are *topical*, because their defining criteria are topical. Every homogeneous area can be analysed into topical elements. Inversely, attention to the regional setting of a topical element broadens the understanding of its connections. *The two are not apart.*

Whittlesey subjected the notion of region to a most profound analysis. He proposed different categories of regions by vertical categorization [single feature region, multiple feature region, total region] and by horizontal categorization [uniform/nodal regions]. I shall deal with these categories later [p. 62] within the framework of a general consideration of the concept of region.

On the whole, Whittlesey rejected region as an objective reality; he saw it more as a device for selecting and studying areal groupings of the complex phenomena found on the Earth. In his opinion, any segment of Earth's surface may be termed a region if it is homogeneous with regard to such an areal grouping. Its homogeneity is determined by criteria formulated for the purpose of sorting out the items required to express or delimit a particular cohesive areal grouping. With this definition, perhaps Whittlesey foreshadows the Regional Sciences, where boundaries and regions are delimited by a pragmatic goal: the region is not an object, but an intellectual concept; it is not self-determined or naturally provided, but an entity for the purpose of thought, created by the selection of certain features that are relevant to an areal problem or interest and by disregard of all features considered irrelevant. This standpoint is a far cry from the landscape approach of the 1930s [p. 41].

Reactions to the quantitative revolution

If, as in Burton's [1963] view, the beginning of the 'quantitative revolution' and the diffusion of the positivist paradigm can be precisely exactly dated, the emergence of new paradigms in geography is more difficult to date. Deviations of and challenges to the positivist paradigm existed from the very beginning of the quantitative revolution [Brookfield, 1964]. As mentioned, the point at which the quantitative revolution reached its peak, was also the period of Hartshorne's elaboration of the regional paradigm [1959]. At the same time that Harvey's 'Explanation in Geography' [1969] was published, a thesis on existentialism in geography [Samuels, Ph.D. 1969] was submitted. On the other hand, the positivist paradigm thrived in the seventies [Chorley and Kennedy, 1971; Bennett and Chorley, 1978] and continues to enjoy a prominent place in the literature today. After the quantitative revolution it was not that a single dominant paradigm came to be applied in geography, but that a number of paradigms emerged and claimed their right to exist. As the task here is to study the regional paradigm, the other paradigms in contemporary geography will not be treated in detail – this has been done by many authors, especially Gregory [1978] and Johnston [1983]. But some attention is due the influences of the various approaches to the idea of regional geography.

What were the reasons and manifestations of the decline of positivist, quantitative geography? According to Gregory [1978], by 1971 this geography was becoming stale. Social involvement in the US and the war in Viet Nam raised the question of why geography had so little relevance to the events of the day. On the other hand, some criticism was levelled against the quantitative methodology rather than the positivist paradigm. Importantly, this criticism came not only from the outside but from within the ranks as well. The criticism of Berry [1980], one of the first and most prominent innovators and protagonists of the quantitative methods [Berry, 1964], was of decisive importance. The spatial field theories had rapidly become 'traditional' statistical geography [Berry 1973], with controversial inferences from statistics and measures of association put to mindless use without regard for the validity of their assumptions. Berry [1971] ridiculed the misuse of quantitative methods as *DIDO* data analysis:

> Data are fed *In* to a large computer with a packaged program in the hope that significant *Dimension will come Out. But it can be that with bad data it will be GIGO – *Garbage In – Garbage Out*.

Opposition to the positivist paradigm was largely based on phenomenological philosophy [p. 164; Marchand,1974; Relph, 1977].* Phenomenology undermined the undoubting confidence of positivism in the certainty of the objective scientific approach to social problems [p. 17]. The assurance that a scientific treating of social and human problems is objective, has general value and is free of subjective influences, was questioned and, indeed, shaken [Brookfield, 1969]. Predating the philosophically conscious reaction to positivism were negative reactions to the positivist paradigm even in its heyday:

– Analysis of spatial processes led to the development of forecasting theories [Haggett et al., 1977]: '... the ability to forecast accurately should represent an ultimate goal of geographical research.' But the prophecies failed, as they took into account only one sector of human activity, the economic- material one, without considering motives and iconographies, the 'hidden factors' [p. 103] which can be discovered only by studying unmeasurable variables and qualities [Sanguin,1981].
– The quantitative approach viewed humanity as a composition of *social groups*. It related not to the *individuum*, but to the society. The behaviour of the individual human being was neglected in favor of 'homo economicus' or 'homo rationalis' behaviour, which could be predicted using statistical probability. In opposition to this rather mechanistic concept emerged the *behaviourial paradigm*

---

* Recall that the term 'phenomenon' was used by Sauer in 1924 [p. 41]. I do not think it likely that his use of the term was linguistic and not philosophical; as the works of the Moravian philosopher Husserl [1859 – 1938], written in German, would have been accessible to Sauer before Husserl's work was translated into English and became familiar to the mainstream of Anglo-Saxon geographers, Sauer was probably aware of Husserl's philosophy.

[Sitwell and Latham, 1979], with its field of study the behaviour of the individual and his relationships to society, environment, place and space [Guelke, 1974; Entrikin, 1976; Saarinen, 1976; Gregory, 1978; Johnston, 1983b].

- The quantitative approach dealt with measurable variables only. But many human qualities which do not lend themselves to being measured are, nevertheless, important motivating factors: emotions, 'Weltanschaung', religion, social consciousness, all motivations which cannot be termed economic or necessarily rational [Russell, 1979]. The need to relate to the human being as such engendered the paradigm of *humanistic geography*. The term 'humanistic' is not new. As a philosophical way of thinking it dates back to at least the Renaissance period [Entrikin, 1976; Ley & Samuels, 1978b]. Good regional geography was always humanistic, but only in the late 1970s did these ideas emerge as an emancipated attitude in the literature [Buttimer, 1978]. Johnston [1983] defined the humanistic approach as being concerned either with the study of individuals or with analysis of landscapes.

Although these behavioural and humanistic approaches brought a renewed interest in human geography as other than an economic or spatial study [Bunting and Guelke, 1979], they did not exhaust the possibilities for directions of research. New paradigms – or, perhaps, slogans – brought to geography a wide range of approaches: practicalism, realism, radicalism, Marxist geography, indeed a pluralism without precedent. Unquestionably, this pluralism is more laudable than the previous regime of large numbers, but it contains the danger of generating *centrifugal forces* which may disperse geography into many directions, in some cases, perhaps, into a way of no return. Pluralism is acceptable so long as the differences among the various geographical interests are not greater than the similarities between them and related sciences [Bartels, 1973]. One cannot deny the very real danger of an explosion of geography if pluralism takes the shape of centrifugal forces unbalanced by centripetal counterforces.

The end of seventies saw a total disorientation of geographical research, as in its disparate subject matters. A *'radical' geography* emerged. According to its proponents, geography should serve the social struggle, be 'relevant' to social problems, take part of decisionmaking. This trend finally became a *marxist* oriented movement [Quaimi, 1982]. Another trend stresses the role of *cultural relativism*, positing even material needs as functions of culture or belief. The term 'man' is not a universal having validity in all circumstances, but each man belongs to a particular group with a distinct culture. The study of the *impact of man on nature*, was dusted off and refocused in the concept of cultural landscape [Relph, 1981].

The ideas of relativism and the landscape school indicate a renewed perception of the environment as seen through a cultural filter. Perhaps we are not so very far from a new regional geography [Turnock, 1967]. More and more geographers now deal with *place* after two decades of virtually ignoring it in geographical studies. Places are being considered as objects to which people relate mentally; attachment to and love of place, *topophilia*, has become an independent object of study [Tuan, 1976]. In French geography, 'région vécue' – a lived region – relates [Frémont, 1976] to landscape and environment as perceived by those who live in it and have a *mental* relationship to it.

All these ideas, which have proliferated over the last ten to fifteen years, need a *focus*, a central point, to which they can be directed, and which possesses the needed centripetal force to unify them within the discipline. As proposed from the first page of this text, the geographical regional approach can unify these ideas both through providing the possibility of further development and by safely anchoring them in geography via the study of places.

Gregory [1978, p. 171], after analysing the state of the art in geography, returns to the merits of regional geography, seeing it as a remedy to the epistemological and methodological crisis in which geography now finds itself:

> Ever since regional geography was declared to be dead – most fervently by those who had never been much good at it anyway – geographers, to their credit, have kept trying to revive it. This is a vital task; we need to know about the constitution of regional social formations, of regional articulations and regional transformations. What makes geography so difficult, is its attempt to operate within specifically regional contexts.

PLURALISM: 'DERNIER CRI' OR 'CUL-DE-SAC'?

Pluralistic trends can reach such a large spectrum of interests that they become no less dangerous than the totalitarian trend of prefering only one approach. Both extremes portend disaster, either that of restricting geography to a rather narrow epistemology ['fermer, c'est de l'enthropie' – 'closing means entropy', Isnard et al., 1981], or that of atomisation into esoteric subject-matters. Nevertheless, a broad, very liberal approach has been proposed:

> There is a place in geography for the poet and the prophet and the seer and the rebel and the critic and even the dogmatic doctrinaire, as they all tramp along the intricate network of often impassable footpaths that now replace the ages-old vision of a royal road to truth [Couclelis and Golledge, 1983].

With all due respect for and support of a pluralistic approach in geography, some definition is required, of either a spectrum or a basis from which geographers will proceed. This standpoint, of course, is not new, but in this difficult period it needs to be stated again. As early as 1924, Sauer expressed the view that '... as long as geographers disagree as to their subject, it will be necessary to seek a *common ground* upon which a general denominator may be stabilised'. Isnard et al. [1981] do not hesitate. The base exists: it is localisation, which is to say 'differentiation' ['... la base existe; localisation, c'est a dire, differentiation'].

Unfortunately, the relationship between the different approaches is not always idyllic; the generation of the seventies was no less critical and dogmatic than that of the fifties. Some scholars seek not a way of reconciliation but a merciless struggle between ideas: 'Each work must be allowed to become the negation of what it was before'[Olsson, 1979]. Nevertheless, Gould and Olsson [1982] did express the need for a *common ground* [not, however, crediting Sauer with the original use of the phrase] within centrifugal trends:

> The strength and the weakness of geography spring from the breadth of its inquiries. In the days as ours of 'partitional thinking', i.e. age of specialisation, a *tradition of integrative inquiry is a value both in the society and university*. On the other hand, fragmented researches, on a large spectrum of inquiry, without structure and connection, are often the issue of it.

What is needed is a focus, a centripetal arrangement of the broad, eclectic spectrum, a 'common ground' to which even eccentric inquiries can be connected. Since the beginning of the eighties, a tendency is recrystalising in geographical thought of concentrating the spectrum of interests in the concept of *place*.

What is the purpose of the criticism against the positivist- quantitative approach? A discipline that lacks goals, lacks order, is not properly termed a discipline [Couclelis & Golledge, 1983]. The focus of the positivist approach was the attainment of levels of explanation of spatially distributed phenomena. What is the aim of the post-positivist schools? Each one has a different aim: the humanistic-idealistic focuses on ideologies and human values; the behaviouristic, on actual behaviour and mental structures; the Marxist, on using geography as a tool in world-wide revolutionary change; the cultural, on studying the structures underlying cultural expressions. I propose that all these approaches be seen as points along a spectrum, expressing the range of *regional foci*, i.e., places from which all these activities, processes and qualities stem, meet, and interrelate to create reality.

I also propose a reconciliation between quantitative and qualitative approaches. Like many other dichotomies, quality and quantity are but extreme positions of a system of values. Contemporary regional geography should put to good use the of models, today indispensable to geographical theory, which were developed by the positivist-quantitative school [Isnard et al., 1981], such as spatial diffusion, migration, urban hierarchies, and so on. The truly valuable legacy of the positivist school – standards of clarity, rigorous development of arguments in the course of inquiry – cannot be ignored by its challengers [Couclelis and Golledge, 1983]. It is perhaps the opinions of reconciliation between the quantitative and qualitative approaches, depending on the particular object of study, which are leading geography in a new direction [French & Racine, 1971].

REGIONAL SCIENCES AND LANDSCAPE ANALYSIS

Along with the attempted discrediting of regional geography by the quantitative and positivistic school, there emerged two new approaches. One was Regional Science, which sought to divorce itself completely from the environmental aspects of geography, aiming for integration into the economic sciences. The second was

Landscape Analysis which stressed mainly the application of geographical research.

Regional sciences

This branch of social sciences, perhaps the newest of them [Isard, 1960], has been thriving for the last thirty years. The closer one's examination of it, however, the more difficulties one notices. A definition of Regional Science does not exist. Isard [1960] gives thirteen definitions, each of which, in his opinion, may be considered valid. Even the definition of the basic term, region – the obscurity of which caused so much of frustration and difficulty in regional geography and was, together with the problem of the definition of regional boundaries, one of the targets of the opponents of regional geography [Grigg, 1967] – is not to be found. In most cases, region is an arbitrary area, serving numeric manipulations. It is strange, ironic and even tragicomic that, whereas the absence of clear definitions of these basic terms caused a massive defection from regional geography, their absence in Regional Sciences does no harm to their popularity and well-being. In my opinion, the appreciation of Regional Sciences in the family of social sciences is anchored in the fact that the Regional Sciences speak a quantitative language, which social sciences scholars consider a language of exact sciences.

It is precisely this exactitude and scientific nature that undermines the validity of Regional Sciences [McDonald, 1966; Mead, 1980]. By neglecting and refusing to deal with elements other than those which are measurable by quantitive methods, it negates an important spectrum of man's social life and his geographical achievements stemming from it: ideology, motivation, devotion, in fact all the ideographic and mental elements, which perhaps cannot be quantitatively measured but which are at the base of human activities. This one-sided approach gave rise to extreme criticism of the Regional Sciences, including even the view that man is an irrational being whose actions and reactions cannot be presumed.

This extreme standpoint is no less dangerous to social studies than that of the Regional Sciences, to which 'homo rationalis' or 'homo economicus' is the prototype of mankind. But even at the dawn of the establishment of the Regional Sciences, Gottmann [1955] warned:

> A country's resources, however, are not all of material nature, nor are they statistically measurable. Any analysis of the quantitatively measurable potential, that neglects the resources of mind and of spirit, will never arrive at a realistic or true picture.

Accordingly, our first reproach to the Regional Sciences is that they deal only with measurable, quantitative elements.

Our second reproach is that the Regional Sciences do not deal at all with the physical challenges man's environment posses to him. Even today, in a period when man, through technological sophistication, can withstand immense natural challenges and hazards, he acts not in a vacuum but in a certain environment.

An example to illustrate this point is Ocean City, in southeastern Maryland on the shores of the Atlantic Ocean [Dilisio, 1983]. This city developed with great vigour over the last thirty years. Seventy-five percent of its buildings are hotels and motels. In 1975 the population of permanent residents was recorded at 4,000 inhabitants; on weekends, however, the population swells to 200,000 inhabitants. There certainly is great interest in studies of its urban structure, tourism, transports, economic developmant, and so forth. But there is in Ocean City another element – the shoreline – which the city fronts at a distance of not more than sixty to seventy meters [approximately two hundred feet]. This shoreline is being eroded at a rate of 0.75 meters per year, which means an advancement of the sea of 82 meters over the course of a century. Thus the essential, vital problem of Ocean City is neither the traffic problems on weekends nor the prices of rooms in the motels; its primary problem is control of this erosion. Men wish to turn each square foot into economic resource, but the ocean continues in its rythmical gnawing away at the continent. There is a possibility of reaching a 'modus vivendi' between the opposite forces, by building sea walls, groynes, even by importing sand to replenish the impoverished shores. But in order to reach a solution, which will be very expensive, one must first perceive that there exists a crucial problem. After finding the right and effective solution, routine efforts can maintain the 'modus vivendi'. Ignoring this environmental challenge, however, makes all other problems of Ocean City irrelevant. Before all else, one must make every effort to prevent the city from being wiped away by ocean. The research within Regional Science does not go in this direction; its subject-matter is economic or socio-economic issues.

With all due respect to the subject-matter of Regional Science, it is no substitute for regional geography. Some authors even accuse Regional Science of misusing the term region [McDonald, 1966]. Today, in fact, Regional Science is a branch of economics, even if some of its scholars are found in departments of geography. The disillusionment with Regional Science may be perceived from these two quotations:

> Arithmetic models designed for posterity may last but for a few weeks [Mead, 1980];

> May we speak as poets?
> May we throw [out] all our equations? [Olsson, 1979].

Landscape analysis

The term 'landscape', 'landschaft', used for generations [Sauer, 1924; Rougerie, 1977], received a status of distinction at the I.G.U. Congress in Tokyo 1980 when a committee for Landscape Analysis was established. I wish to examine whether the term 'landscape' may be substituted for the term 'region', and whether landscape analysis can inherit the place of regional geography.

The terms we use originate in common language and have a rather general meaning. From them we try to create, by limiting their meaning, a 'terminus technicus'. The term is sent to the world on the pages of our publications and on the wings of our lectures. But a term has a life of its own, and as it wanders the world, its meaning changes to become something other than what was intended at baptism. This was the fate of the term 'region' and certainly of the term 'landscape'. Landscape is used not only by geographers, but also by artists – painters, poets, novelists – and by architects, to whom 'landscaping' means to embellish a certain area, usually derelict land, sides of highways or building plots, into gardens and parks [Brassard and Wieber, 1984]. In Schools of Architecture there are Departments of Landscape Architecture; engineers are taught Landscape Engineering. It is very difficult to prevent ambiguity and misunderstanding when the basic term is understood differently by different disciplines. It seems that with regard to definition this term is even more complex than the term region.

The study of landscape is deeply anchored in the geographical tradition and currently receives considerable attention from many scholars [Loewenthal and Prince, 1964, 1965; Birdsall and Florin, 1978]. Generally speaking, natural and cultural landscapes are differentiated [Meinig, 1979]. The importance of the study of landscape is often in revealing the culture of its population or past populations [Bertrand, 1984]; hence the importance of the study of landscape in regional as well as historical geography.

Geographers dealing with landscape see as their main subject the *changes from natural to cultural landscape* [Neef, 1967]. From this standpoint, landscape analysis has great importance in ecology and planning. It also has great pragmatic value [Děmek, 1978], and perhaps it is one of the branches that make geography an applied science [or which causes geography to be applied]. It is not surprising that landscape analysis is highly developed in Eastern and Central Europe, where planning is one of the building stones of political action [Mazúr, 1983; Mazúr, Urbánek, 1983; Drdoš, 1983].

On the other hand, a restriction of landscape analysis to the study of the natural environment and the activity of man in changing natural landscape into cultural landscape leads to a limitation of its epistemology; there are only a few studies by landscape analysts in the urban realm. It seems to me that in spite of its importance and its affinity to regional study, dealing with landscapes cannot substitute for dealing with regions.

REGIONS, ZONES, BOUNDARIES

Men and women are not only themselves; they are also the region in which they were born.

Somerset Maugham, *The Razor's Edge*

Geographers seek regions as alpinists seek unconquered mountains: 'because they are there'.

J.R.McDonald, 1966

CONCEPTS OF REGION

Regional geography is a very difficult discipline. It has no 'instant' experts [Hart, 1982]. On the contrary, regional geography requires 'sensitive interpretation of the values that motivate the behavior of the people, which is a *lifetime pursuit*' [Hart, supra cit.]. Good regional geography must be grounded in keen sensitivity to the needs, wishes and values of the people who live in the region.

The difficulties and discrepancies of the definition of the term 'region' can be summarised in two opposing postulates:

- Region is only a mental concept, a tool for study, a 'model' elaborated by a scholar to accomplish his study;
- Region is a reality, existing in space; the scholar's task is simply to study it.

James' definition [James, 1959b] represents the first concept:

> ... region is an area of any size that is homogeneous in terms of certain criteria and that is distinguished from bordering areas by a particular kind of association of areally related features. The region is a device for illuminating the factors of a problem which otherwise would be less clearly understood. *It is not an objective fact; rather it is an intellectual concept.* There are as many regional systems as there are problems worth studying by geographical methods.'

Beaujeau-Garnier et al. [1979] also see in regional study a global study of a piece of territory, arbitrarily delimited ['... l'étude régionale d'un morceau du territoire qu'on découpe comme on le veut']. Edwards [1970] considers the region to be a purely intellectual concept: as the historian uses a device to assist him within the continuity of time, namely, a *period*, so the geographer, dealing with the continuity of space, uses a device named *region*. To this I would reply that time and space are not of the same substance; delimitation of time is necessarily intellectual, whereas delimitation of space can be according to tangible elements [Minshull, 1962].

Others have it that region is a reality: '... les régions existent en dehors de chercheurs; ils les doivent decouvrir, pas de les créer' ['regions exist outside the scholars; they must discover them, not create them', Dumolard, 1980]. According to this concept, a region results from a cutting-out of first order by a state-nation [Langton, 1984]. The concept of region denotes [Reynaud 1982] both relationships between a social group and a certain area, and relationships that this social group establishes with other social groups, which also have relationships with the areas in which they are situated [Daude, 1971]. Pred [1984] sees in settled places and regions the essence of human geographical inquiry. Regional units are seen [Harper & Schmudde, 1984] not as self-contained entities but as *interacting components of national or global systems*. These authors see the geographical task to be solving the 'geographical equation':

L [life and space] = E [environment] · C [culture],

where culture is everything acquainted with the mind of man [religion, behavior, ideology]. Spatial interrelations are the connections between the place under study and other places. The solution of the equation proceeds by use of four variables: examination of the physical, cultural and economic geography of a place, and understanding the interactions of these elements. This approach is just as pragmatic as the approach considering the region to be only a tool: *the region is the principe of management of the territory* [Juillard, 1962].

Perhaps a way to bridge the opposite concepts of region – reality versus mental construct – is the idea expressed by Claval [Beaujeau-Garnier et al., 1979]: the concept of region is universal, but the type of region varies as a function both of history and of the country's evolution.

I propose to consider region as both reality and mental structure. I will defend this proposition by means of an illustration, Megalopolis. The urban structure of the area between Boston and Washington, D.C. and the relationships between its different components existed well before 1961, when Gottmann revealed the concept of Megalopolis – a composite structure of urban, suburban and rural elements interacting in a certain mode of relationships. Megalopolis is a reality, which cannot be denied; on the other hand, the internal structure of this reality was not perceived until it was defined and formalised in Jean Gottmann's mind. We must accept Megalopolis as both a reality and a mental structure. This is not a contradiction: it is the human mind that discovers the processes hidden behind real phenomena.

In most regional researches, the region is a reality, and the role of mental activity is to formalize the structure behind this reality. Laity [1984] considers region to be both mental structure and concrete object. He improves the current definition of a concrete object as something which has a spatial extension and is independent of thought. His definition is that a concrete object is a union of specified elements which are considered associated for the purpose of the observer.

The idea of a region being both reality and mental concept has to do with the 'total' region [which I propose to label 'systemic region' [p. 66]; 'functional' regions [p. 62] having a certain pragmatic purposal, exist only in the mind of the scholar, politician or other interested person. Such arbitrary regions 'merely enclose space and resemble trash cans' [Ullman, 1954]. Thus in answer to the question of whether a region is a tool or a goal, I would state as follows: the systemic region is a goal, being the structure which generates processes that we are eager to understand; the 'functional' region is a tool, a framework for a certain, pre-defined purpose.

The two aspects – goal or tool – also express the difference between the more philosophical and the more practical interests of scholars. The approach of the Regional Sciences is of a practical nature, considering as a region an area homogeneous with respect to certain announced criteria.

Regions exist; they need only be discovered [Frémont, 1976]. A region is an area lived, perceived and experienced by a human being. The concept of region has to take into account that the importance of a *place in space* grows with the *polarisation* of activities on that place [Frémont, supra cit.]. Most poles are cities. A pole has three functions: *a.* executing changes between the place and its surroundings and the exchange of goods, services, information, capital; *b.* being a place of social interrelationships; *c.* being the center of growth. Continuing polarisation of a place creates from it and from its surroundings a region; environmental factors such as the physical background, topography, hydrography, and so on are also of special importance. Growth is not schematic; specialisation of each place in a certain branch – industry, tourism, higher education, administration, and so on – can modify its growth. Antecedent historical structures have their particular influence and can lead growth in a particular direction unless the pole is situated in a new country, in a 'tabula rasa'.

Region is a product of a society [Isnard, 1978].

Every region has its heritage of good and its burden of evil. Every inhabitant from child to patriarch should strive to know what his region contains, not only its wealth of natural resources, scenic beauty, and heritage of culture but the opposite picture as well: the evils of ugliness, poverty, crime and injustice of all kinds. The citizen must first study all these things with the utmost realism, and then seek to preserve the good and abate the evil with the utmost idealism [P. Geddes, 1915].

As I stressed elsewhere [p. 12], regional studies are favorable to conservation of regional individualism, the products of history and environment [Sorre, 1948]. To understand a region, we must understand the different ethnic groups living within it [Frémont, 1980]. In the same area, each ethnic group has its 'espace vécu', its lived space. The author, having lived in Jerusalem for forty years, can certainly agree with this idea: Jerusalem as the lived place of an Orthodox Jew is not at all the same as that of a Greek monk or a Muslim merchant; neither is the life of a secular Jewish low-income worker the same as that of a successful industrialist. A regional study has to answer the problems of social groups, but should not neglect the actual relationship between the individual and the region [Wright, 1966; Loewenthal, 1961].

DEFINITIONS AND TYPES OF REGIONS AND ZONES

One of the difficulties in dealing with regions is the multitude of definitions of the term 'region'. The cohort of definitions stems from the variety of mental images of the region held by different scholars. Authors denominate by the term 'region' both unique and repetitive realms. I propose to define the different realms in clear terms: *region means the unique, zone means the repetitive realm*, even if the relationship between the two is flexible. There are cases in which a zone becomes a region: coastal zone is a repetitive environment, existing in many parts of the world. But a settled coastal zone is a part of the Ecumene and belongs to a certain nation, a certain district, and is characterized, besides having the repetitive qualities of a coastal zone, by the unique qualities of the particular district. The use of the terms region and zone should be consistent, using the basic differentiation between the different areal realms as proposed by Whittlesey [1954], Hartshorne [1959] and Berry [1964, 1968]. These relate to the number and nature of the variables in the realm, which may be characterized by one, two or more common variables. There are four main areal realms, differentiated by the criterion of the occurence of common variables:

- a. The *functional region*, defined ad hoc for a certain purpose, task or aim.
- b. The *single feature region* [Whitttlesey, 1954] or *uniform region* [Berry, 1964], a geographical realm characterized by a single common variable.
- c. The *multiple feature region*, a geographical realm characterized by two or more common variables.
- d. The *total region*, a geographical realm characterized by the sum of man's activity in it; I propose to label it '*systemic region*'.

a. Functional region

This is a pragmatic delimitation of a certain area in order to fulfill a certain aim, with no ambition to see in its boundaries more than ad hoc limits of a certain interest. A priori, its boundaries and the variables to be treated have been choosen and defined, and there is no 'raison d'etre' for it other than the execution of the task for which this framework was established. Therefore, neither a theoretical consideration of the nature of its boundaries nor consideration of its content as an organic unity is necessary; its existence begins and ends with the task and purpose which delimit this area from its surroundings. The task may be of a scientific, political or economic nature, such as delimitation of an areal study by a certain budget, delimitation of an electoral district, delimitation of a development scheme, and so on. This type of region is extensively used by the Regional Sciences, but is not a 'region' with all the attributes inherent to this term [cf. p. 55].

b. Geographical realm defined by one common variable

Of all geographical realms, the boundaries of this one are the most exact: the distribution of the common variable is both the content and the limits of it. In fact, the realm of one common variable is the spatial treating of one of the elements of topical geography. It can be applied to a variable such as a slope, expressing the distribution of a slope of a certain degree, or as population, expressing the distribution of a certain quality of it, in each case the variable having been selected and defined a priori. This type of geographical realm is both *uniform and homogeneous*, as it contains *one* variable distributed *continuously* over a certain area. This type of geographical realm is repetitive; slopes of a certain degree and population of a certain density can occur at different places. This type of realm is not unique, as the area under study is only a segment of a global area of this variable. I propose to call it 'zone', which means a *homogeneous, uniform, repetitive realm of a certain value of a variable*.

There can exist, on the other hand, a geographical realm possessing a common variable the distribution of which is not continuous, with no areal homogeneity, existing only as a certain function between points, or between a focus [or foci] and points. This function can be something such as commuting [Fig. 2] between residence and work-place, links between the central bus station and the terminals, distribution of certain services or goods, or any activity, the distribution of which is not necessarily homogeneous. In this case, even if we are dealing with only one variable, the relationship between its parts at a given moment is unique, changing over the time. Because of its uniqueness, the relationships being not necessarily repetitive, this type of realm is a region, and in this special case, a *nodal region* [Boudeville, 1968b; Minshull, 1962].

A nodal structure can be attributed also to a functional region, which is defined and delimited by activities originating in a focus or pole. One example is distribution of regional – not national – newspapers, which shows the close connection between a central place and its market area [Blotenvogel, 1984]. The distribution of regional subscription newspapers in East Westfalen [West Germany] for the year 1982 has been plotted in Fig. 3. Even if the distribution is nodal, and despite overlapping, it can be seen that in most cases a regional newspaper is limited to a certain spatial diffusion.

c. Geographical realm defined by two [or more] common variables

Two types of distribution, the homogeneous and the nodal, are discernible in this realm. The variables can either be independent of each other or derive from the same source. If the combinations between the variables are repetitive – as different elements of climate, land use, business centers, industrial realms, and so forth – then these realms are zones and not regions. Socio-economic zones, even if repetitive, become unique when blended with the local qualities: the Ruhrgebiet is certainly part of a repetitive industrial zone, but at the same time it is a unique form of organization and functioning, and there are no two Ruhrgebiete in the world. The Donbas industrial zone can be compared but not identified with

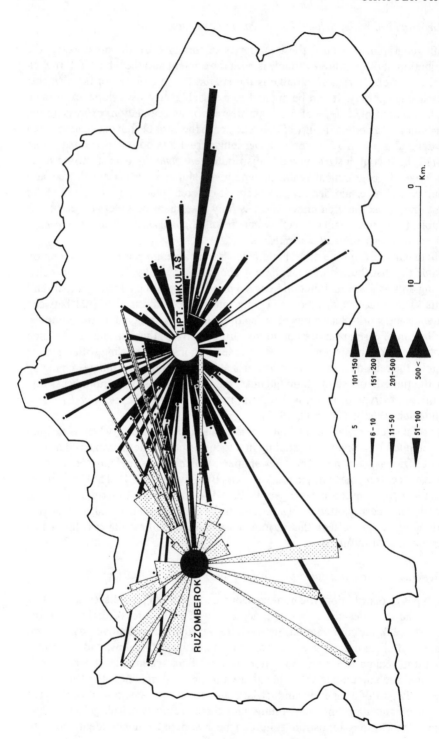

Fig. 2. Daily commuting to centers of employment as a nodal organisation of a region, the towns of Liptovský Mikuláš and Ružomberok in northern Slovak Soc. Rep. [Sipka, 1970].

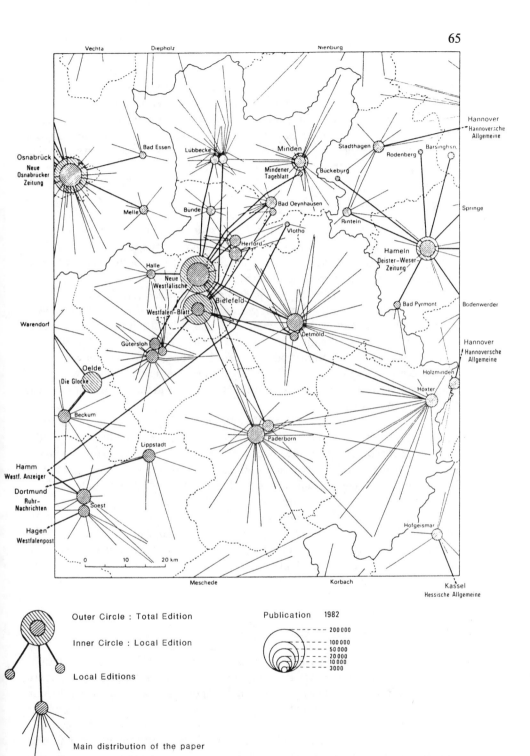

Fig. 3. Distribution of local newspapers as a regional function concentrated in foci: East Westfalen, G.F.R. [Blotenvogel, 1984].

Ruhrgebiet's uniqueness, because it possesses in addition to the repetitive qualities of an industrial zone its special local qualities. Therefore, a geographical realm of two or more variables can be either a zone, which is repetitive, or a region, which is a unique phenomenon. The choice of qualities to which one pays attention depends on the aim of the research: to the repetitive ones, relating the area to a world-wide zone [climatic, agricultural, industrial], or to the local ones, lending to the repetitive area the characteristics of a region.

d. 'Systemic region' ['total region']

The breakdown between zones and regions depends not on the number of common variables, but on the existence or absence of interactions, relationships and processes between the variables. Uniqueness, the core of the discord among geographers, exists only in the systemic region, which we define as a *certain part of the earth's surface, where the population, organized within certain social, political or administrative boundaries, faces natural, social, political and economic challenges.* Regional research studies the processes in action at a certain time. What makes the region an integrated entity [Martin and Nonn, 1980], a 'totality', is the sum of the relationships and interactions between the acting elements in the region. By this definition, a region is characterized by the processes acting in it.If we accept this definition, we will see that the concept of a region and the concept of a system [p. 75] are the same.

   This type of region was termed [Whittlesey, 1954] 'total' region, as being characterized not by certain a priori defined variables, but by the totality of the variables incorporated in the area and the interactions between them. The adjective 'total' was misunderstood, as if each element existing in the area should be incorporated and investigated; in fact, the intention is to include only to those elements which are relevant to the structure of the region.

   This type of region fits the concept of 'pays' of the Vidalian school [p. 38]. An attempt was made to denominate this type of region 'compage' [Whittlesey, 1954], but this did not find wide acceptance. I propose that this type of region be termed 'systemic region', as the systems approach fits both its epistemological and methodological postulates.

HOMOGENEOUS AND NODAL REGIONS

It is quite difficult to imagine a region, with homogeneous socio-economic, cultural and physical qualities, over a surface of tens or even hundreds of square kilometers. But homogeneity should not be understood only as a continuous distribution of one element, but also as a homogeneous distribution of different elements. Take as an example a middle-sized farm in an agricultural region, the Paris Basin, for instance. In such an average farm we shall find certain cultivations: sugarbeet, wheat, pasture and orchards. This type of land-use will also be found in the surrounding farms in the Paris Basin. Thus, despite the variability of cultivations in particular farms, there is a certain homogeneity in the land use on a regional scale [Ivanička, 1980]. In the same way we can speak of the homogeneity of the urban space, not that every street in the city is uniform, but that there is a repetition of the same elements [residential quarters, industrial zones, parks, shopping centers, schools, garages, etc.]. *Homogeneity of a region is but a distribution, organized in a certain way, of the elements that constitute the region.*

Poles of growth are the basis for defining the 'nodal' region. As the city becomes more and more important by its concentration of population and its functioning as a focus acting on its surroundings, the perception of a nodal [or polarised] region becomes more and more important in regional geography [Brunet, 1972a,b]. In fact, urban geography can be considered regional geography, if we considering the city as a homogeneous or even nodal region, because its functions extend beyond the municipal area [Frémont, 1976].

The model of the nodal region was a welcome innovation, capable of over-coming the criticisms of the homogeneous or total region. In the nodal model the boundaries become secondary, and in fact can be ignored altogether, as the connections between the focus and its periphery [Fig. 2] are based not on spatial boundaries but on the intensity of relationship and interaction between them [Dickinson, 1970].

I would argue that the nodal region, at least in some cases, is only a stage in the development of a systemic region. I propose a model showing that total regions of today developed, by historical process, from nodal regions.

This *model of a systemic region developed from a nodal one* is illustrated by suburban Maryland, close to Washington, D.C. During the 1920s this part of the state, between Baltimore and Washington, had the distinctive pattern of a nodal region [Fig. 4.B], based on a road system that led almost exclusively to and from the focus of Washington, from the northwest, north and northeast to the south. There were no east-west arteries, with the exception of one in the northern part leading to Baltimore. In the late twenties along these southbound roads – Georgia Avenue, Old Georgetown Road, Rockville Pike – there developed, together with older settlements, small towns such as Rockville, Silver Spring, Norbeck, Glenmont, Wheaton, Kensington, Garret Park, Halpine. The region was clearly a nodal one, with a sharp distinction between the focus [or pole] and the periphery. The function of the region was that of a residential area, with downtown Washington being the center of commuting. Even twenty years later, in 1949, notwithstanding the great development of these towns, the nodal structure is still dominant [Fig. 4.C]: roads in the direction east-west are rare, and the tendency is still to attain the focus. Between the main arteries leading to Washington there are still many unbuilt areas, the towns are rather isolated, and no homogeneous urban landscape [in the meaning expressed above] can be traced.

After another thirty years, by 1980 [Fig. 4.D], the nodal structure has become a homogeneous one. In addition to the five main highways in the north-south direction [one of them an Interstate freeway], three highways travelling east-west, one of these the Beltway, have been constructed. Most of the area between the main highways has been built up, with the topographically difficult ground around

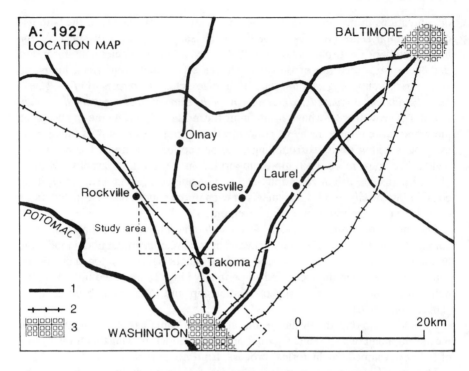

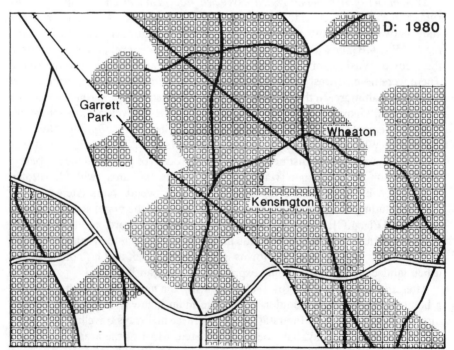

Fig. 4. Transformation of a nodal into a systemic region, Suburban Maryland, 1927–1980.

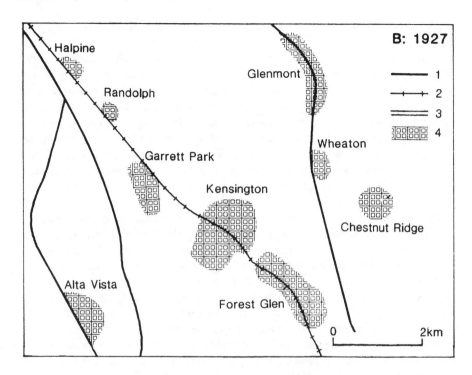

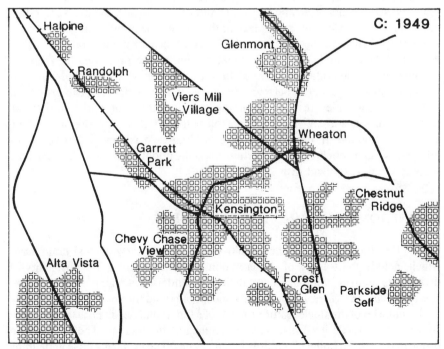

1. Routes.   2. Railways.   3. Built-up area.

Rock Creek left as a park, i.e., having an urban function as well. The changing of a nodal into a homogeneous region, one in which occurs a certain rhythmical repetition of residential areas, shopping centers, industries, schools, garages and parks, is not only a matter of building up the area but also of introducing into it most of the urban functions which characterize a city, so that the difference between focus and periphery diminishes. Near the important crossroads have been established important, mostly white-collar industries. The movement of the population is not only the daily commuting to downtown Washington as a focus of employment, but also to important foci of employment in the suburban area itself. There is even a movement in the opposite direction: these are the residents of Washington commuting to their sites of employment in suburbia. More and more the possibility exists that a resident of the suburban area will not reach downtown Washington for months [R.A. Harper, personal communication, 1985] because the suburban area provides not only his place of residence, but also employment and all the essential medical, administrative and cultural services. The nodality of downtown Washington is no longer a critical value.

If this development is accepted as a certain model of behavior, we can argue that under conditions and processes favorable to it, a nodal region develops into a systemic region. On this basis we must inquire whether regions defined as such – Burgundy, the Midlands, the Tyrol, as random examples – which have many common variables in land use, industry, traditions and the social interweaving of the population [Langton, 1984] were not at a certain historical moment only nodal regions. At a given period the focus of a region was the feudal castle, linked to scattered settlements by erratic economic and political relationships. At a certain point in history, the city took the place of the castle. The population grew and the settlements themselves became secondary foci for the new population. The region developed from a nodal to a homogeneous one, as the processes between the components of the nodal network led to a 'totalisation' of the relationships and to homogeneity in the area between the focus and its periphery. The growth of West European cities resulted from the transformation of nodal into 'total' regions: Paris absorbed villages like Montmartre, Passy, and the same thing occured in London. In our time the case of Greater Washington illustrates this process. We can conclude that in certain cases the nodal region is an intermediate stage in the development of a systemic region.

DEFINITIONS OF BOUNDARIES AND LIMITS

Grigg [1967] argued that as the 'total' region had no clearly defined boundaries which include the totality of the common variables, there is no possibility of defining its content: without clearly defined boundaries, a region does not exist. It is necessary, therefore, to clarify the concept of boundaries of a region. Defining the boundaries of the single feature zone or of a functional region poses no problem [p. 63]; the variable itself also sets the applicable boundaries. There is also no problem with the defining of multiple feature zones. The problems arise in defining the systemic region.

In literature we meet a whole spectrum of 'boundaries': natural boundaries, historical boundaries, political boundaries, administrative boundaries, etc. [Frémont, 1976]. We must exclude from a regional study the notion of 'natural' boundaries [cf. p. 33]. *Natural boundaries exist only between natural elements*, and not between socio-political and economic elements, which, together with environmental elements, constitute regions. A natural boundary exists between mountain and plain, between ocean and continent, between valley and river. Boundaries between socio-political entities, such as states and countries, are maintained not by natural phenomena, but by the economic, political or military force of the neighboring states and countries. Of course, in the process of fixing boundaries between states and countries there are arguments of accessibility, of possibility of movement, of defense, of economic entities, etc., and thus there is an understandable tendency to base boundaries on tangible natural features in the landscape. It is not accidental that even today there is a high correlation between rivers and political boundaries, as a heritage of days past when rivers were visible, clear and relatively steady landmarks [the Potomac, the Rubicon; Gottmann, 1980]. On the other hand, just as many boundaries have no correlation with landscape elements, as they were established against a background of political and military disputes. In most cases, so-called 'natural' boundaries are only tools of politicians who, interested in some territory, built on that irrational, atavistic argument that a political boundary should have a 'natural' base. Of course a river in a wooded area is a distinct landmark, especially when the area is not developed; but by the same argument a river in such a situation is more a means of transport, of connection, than a boundary which by definition limits contacts. Even if in the past landscape elements influenced the dispersion of social and ethnic groups, today these groups are the political forces which create political boundaries between countries. Another concept similarly misused is that of the 'historical' boundary, [p. 22].

Administrative and political boundaries differ among populations, economic structures and legal systems, and they create different geographical facts on both sides of the boundary, even if the environmental background is similar. A glance at the boundary between California and Mexico [p. 120] or between Montana and Saskatchewan illustrates this. Administrative boundaries between counties, districts and other territories smaller than states are a great aid in regional study, as they are also statistical entities where data is collected. But such a statistical boundary is not a must and its absence can be overcome.

The way to define the boundaries of a systemic region is by identifying and delimiting the extension of the interrelations and the processes acting in the region: *the area where the interrelations and processes fade out and cease to exist, and another spatial system of relationships begins, is the limit of the region.* Depending on the characteristics of the region, this boundary may be administrative, political or otherwise described. In any case, the bounded area must contain a distribution of the essential variables which establish the function of the region as a system.

If we accept the concept of the systemic region, then its topographical boundaries are constituted by the limits of its local input [p. 91] and by its landscape elements, as well as by the limits of its regional output; but, being an open system, the national and global input of the region, as well as its national and global output, are not contained within the topographical boundaries of the region.

REGION AND LITERATURE

Can literature, by faithful description of a landscape, a region, a city, or even a part of any of them, fulfill a role in regional geography as a reliable source? Can literary writing give to a geographical description more authenticity? Conversely, can regional geography enrich literary writing? As a first testimony, let us quote a writer:

> ... Literature begins with a place. A story or a novel which lacks a strong sense of place, absorbing particularities of appearances, of trees and meadows, buildings and streets, objects and tools, odours, murmurs – is not 'truthful' in our eyes, it seems fallacious, a detestable attempt at errecting a curtain with no reality behind it. We forget many of the plots described in the books we read, but the appearance of the places, and in them the people- rooted-on-the-place, is deeply anchored in our memory and is not erased [Aharon Meged, 'Haaretz', 14 March 1986; Hebrew, author's translation].

The relationship between literature and geography has recently been treated by Lutwack [1984] and Mallory-Simpson [1987]. On this subject Tuan [1978] makes three assumptions:

a. Geographical writing should have greater literary quality;
b. Literature is a source material for regional geography;
c. Literature provides a perspective on how people experience their world.

As for [a], Tuan assumes that literature in geography is merely decoration; geography's concern with art has been very slight, and the geographical description, even if it is persuasive and colorful, deals only with external facts and features, not with the inner world of feelings, intentions and so on. This is not always the case, however. Many of the geographical descriptions in travel books and travel diaries, on the border between geography and literature, between science and art, are more than factual descriptions.

Concerning, literature as a source for geographical material abounds, beginning with the Odyssey and the Bible, to Cervantes, Dickens, Upton Sinclair, to name a very few of the possible examples.

As to [c], the regional novel as a literary form appeared in France and was quite common in Britain [Darby, 1948], as witness the works of Sir Walter Scott, Thomas Hardy and so on. From some of these works the actual places can be identified, as, for example, Dorchester described under the name of Casterbridge, and Lafayette County clearly decipherable from Faulkner's descriptions of Yoknapatawpha County [Aiken, 1977, 1981].

The writer does have a sense of place. This is more true of the traditional than the modern writer, however: '.. a traditional storyteller fixes listeners in an unchanging landscape combined of myth and reality. People and place are inseparable'[Louise Erdrich, N.Y. Times Book Rev., June 24, 1985]. And she continues:

> A writer must have a place... A place to love and to be irritated with... Through the close study of a place, its people and characters, its crops, products, paranoias, dialects and failures, we come closer to our own reality. It is difficult to impose a story and a plot on a place. But truly knowing a place provides he links between details and meaning. Location, whether to abandon it or draw it sharply, is where we start.

I do not argue that regional geography and literature are interchangeable. At the same time, however, if we are to make a worthy effort to understand a region, we should neither renounce literature, a priori, as a means to engender understanding, nor should we deprive our text of the metaphors and descriptions that can convey the specific feelings a certain landscape or region can generate.

The concern with the relationship between regional geography and literature leads to an examination of the relationship between regional geography and culture. Culture is [Norton, 1984]:

> ... [the] sum total of human learned behavior and ways of doing things. Culture is invented, carried on, and slowly modified by people living and working in groups, as each group occupies a particular region of the world and develops its own special and distinctive system of culture.

Where is the meeting between culture and regional geography? Acording to Norton, there is a regional background to the systems of culture. As culture in its broad meaning is omnipresent in human behavior and acting, it is an integral part of the approach to regional geography. To formalize this idea: regional geography studies the *impact of culture on landscape*. Some schools in geography consider the transition from natural to cultural landscape as the central issue of geography [p. 57].

We have no choice but to bridge the concept of regional geography as a social science and the concept of cultural geography. In a systemic region, which tolerates subsystems as a part of its stucture, cultural and social as well as natural subsystems can exist and function, sharing the common ground of a definitive location.

REGION AS A SYSTEM

Aucun fait n'est etudié pour lui-même, mais comme partie d'un tout, rouage d'une grande machine. [No fact is studied for its own sake, but as a part of a whole, as a cog in a great machine (author's trans.)]

E. De Martonne, 1902

General system theory represents an exciting and challenging routeway, not just to the reunification of geography but to a new unified understanding of the unity of nature, science and society.

Martin J. Haigh, 1985

BASIC REMARKS ON THE SYSTEMS APPROACH

Why the systems approach?

General System Theory was first proposed as an abstract reunifying field of study in the life sciences [Von Bertalanffy, 1951, 1962, 1969, 1971].Since then, the systems approach has become widely used as well for nonscientific purposes, as in technology, industry, communications, commerce, and so on. The systems approach became necessary in view of the increasing complexity and development of modern science, technology and society.

The real world is immensely complex. A study of isolated parts of reality is practically a decomposition of the real world [Chorley and Kennedy, 1971]. The real world is continuous, but isolated structures are also portions of reality. The road to understanding requires in some sense the breaking down of the real world into meaningful sections. Each section should be both sufficiently complex to possess a degree of internal coherence and sufficiently simple to allow for comprehension and investigation. Systems can answer these prerogatives: they can be identified at all scales of magnitude and with all degrees of complexity. When put to proper use, the systems approach can promote understanding of societal interrelationships. It challenges traditional scientific thinking in the direction of revising the mechanistic, reductionist approach which leads to the atomisation of science.

We have neither the intention nor the competence to treat systems theory extensively. I do wish, however, to briefly review the idea of the systems approach and its essential structure with a view to its use regional geography; in the next chapter, I will elaborate a model of systems approach to this subject. Certain ideas concerning region and the systems approach, such as the concept of holon which is basic to my approach, have already been treated [pp. 23–24].

Why the systems approach? For every field of knowledge [Bertalanffy, 1971], from physics and biology to the behavioral and social sciences, we must deal with its comlexity in order to understand it thoroughly; we must consider the element as a component of a larger 'whole', a system. From a pragmatic point of view, alternative solutions to a certain problem must be considered, and those promising optimisation, maximum efficiency with minimal cost in an extremely complex network of interactions, must be selected. At the very outset the problem of mathematisation of the system study emerges. Most scholars are persuaded that systems can be treated only through mathematical procedures. But Bertalanffy argues [1971, p. 34] that

> there are fundamental problems for which no mathematical techniques are available

and

> ... a verbal model is better than no model at all. Theories of enormous influence such as psychological ones are unmathematical.

The central idea of the systems approach is 'wholeness'. Phenomena cannot be understood by investigating their respective parts in isolation. The concept of wholeness can serve existing tendencies in some natural and social sciences toward integration, and which is *inherent* to regional geography. The systems approach is important in scientific education. Science must have its generalists [Bertalanffy, 1971]: '... there is a need for simpler, more unified approach to scientific problems. Any research group needs a generalist'.

On the other hand, Bertalanffy does stress the importance of the individual component within the system:

> Man is, before and above all, an individualist. The real values of humanity are not those which it shares with biological entities as the function of an organism or a community of animals, but those which stem from the individual mind.

Structure of a system

My reasoning on the structure of a system and its use in social sciences and particularly in regional geography is based on ideas developed by Bertalanffy [1971], Koestler [1968], Chorley and Kennedy [1971], Bennet and Chorley [1978], Miller [1978], Boguslaw [1981], Wilson [1980, 1981] and Huggett [1980]. The reader is recommended to these sources. Here I will mention only the most basic concepts necessary for advancing further toward the application of the systems approach in regional geography.

The term system is employed across a broad range, the 'solar system', for example, or 'Union Pacific System' [Miller, 1978]. Different people mean different things by it. One of the simplest and most concise definitions of the systems approach is given by Wilson [1981]: '... *an object of study which is made of a number of interrelated components*'. On this basis most objects of study are defined as a system. But an emphasis on analytical study restricts the issue to complicated systems, whose components exhibit a high degree of interdependance.

A general definition is that *a system is a set of interacting units with relationships between them*. This implies that the units have some properties in common; if the units are to interact, it is essential that they become *components*. An important aspect of a system is that the *state of each component is constrained, conditioned and dependent on the state of the other units*.

There are three objectives when handling complicated systems [Wilson, 1981]:

a. to set up frameworks to handle the complexity: *description* of the object of study; *structure* associated with it [= its components]; *processes* [= the way in which the parts become a whole].
b. to identify *systemic behavior*.
c. to seek *generality* [methods of analysis in one system which can be used in others].

A system can be *conceptual*, when its units are words or numbers, or *concrete*: a non-random accumulation of matter and/or energy in a physical space-time [= region], which is organised into interacting, interrelated subsystems or components [Miller, 1978]. All change over time of matter and/or energy and/or information [knowledge] in a system is a *process*. The arrangement of a concrete system's parts, at a given moment, is its *structure*. The boundaries between structure and process are not constant but are subject to *change*. Processes may change the structure of a system and, in so doing, change the entire system. In order to survive, the system must adjust to the changes in its characteristics, by a mechanism of reactions to them, *feedbacks*.

In most definitions of a system appears the statement that '*a system is more than the sum of its components*'. This may be misleading: there is no new factor besides the components; *the 'more' is the relationship created by the behaviour of the components* [correlations, interdependence, harmony].

Functioning of a system

Chorley and Kennedy [1971] divide systems according to their functioning into isolated, closed and open systems:

Isolated system – boundaries closed to the import [= input] and to the export [= output] of both matter and energy. These occur more in the laboratory than in reality.

Closed system – boundaries that prevent input and output of matter, but not of energy, as, for example, the Earth.

Open system – exchanging matter and energy with its environment, where a steady output of matter and energy is balanced by the input; the systems and their components tend to become adjusted [= self-regulated to produce a steady state]. The majority of the natural and the socio-economic systems are open systems.

Chorley and Kennedy, and Bennet and Chorley [1978] deal extensively with the structure of systems, and again the reader is recommended to these sources.

Nevertheless, I will mention here certain definitions which are essential for the reader of this text.

> *Structure* of a system is its organisation, or the interrelation of its components. The size and number of variables of a system are its *phase space*. *Correlation* measures the strength of the linkage between the components.
>
> *Input* is the *flow* of matter, energy and information [which may take the form of capital, knowledge, etc.] in the system. It can be *internal*, endemic to the system, or *external*, entering the system from its surroundings.
>
> *Output* can be either the actual export from the system of matter, energy and information produced by the processes with in it, or changes induced to its form and structure.

One of the most important ideas of the systems approach is *feedback* or change in the system introduced by one of its variables. Feedback can be negative or positive. The most common type of feedback is *negative*: an externally produced variation initiates a change which has the effect of damping down, or *stabilising*, the effect of the original change [i.e., *conserving the structure of the system*]. This is a form of *self-regulation* which promotes the open system, or its *dynamic equilibrium*. *Positive* feedback reinforces the effect of externally induced change, and can eventually change the function or the structure of the whole system. Its structure has a built-in self-destructive element.

Negative feedback is vital to the structure of process- response systems. They bring forces into play which tend to oppose continual change in the system caused by a variation input. Feedback mechanismes are constantly at work, tending to bring about balance or equilibrium between the components. Equilibria can be static, stable, unstable, metastable or in a *steady state*, where the interacting variables oscillate within a certain range. Steady state means that the system's components are in motion [Wilson, 1981] but the motion is constant; this is the dynamic equilibrium. To reach this dynamic equilibrium, the system needs a certain *relaxation time*. The intermediate stage between disturbance and equilibrium is the *transient time*.

A system is represented by *system diagrams*, using defined graphic forms. Storages of variables are designated by rectangles; related features, such as *regulators*, by ovals; the *flow* of information on variables, capital, etc., by arrows; feedback, as arrows in the opposite direction; subsystems by discontinuous rectangles, etc.

SYSTEMS APPROACH IN GEOGRAPHY

> *The words are new, but the mental image of such structures of interconnected parts goes back at least to the great Philosophers.*

> P. James, 1972, p. 511.

Some geographers, or generations of geographers, are more interested in individual themes, while others are interested in *complexes of phenomena* [Zonneveld,1983], which can be considered systems. Haigh [1985] identifies at least four traditions of systems thinking in geography:

1) *Physical geography*, although long since divided into its special fields [geomorphology, climatology, biogeography, etc.] is still considered an integrated field of interests. Actual geomorphological processes are studied together with climatic and biogeographical controls [Yair, 1983]. An excellent example of this approach is the 'natural regions of the world' ['Les régions naturelles du globe', Pierre Birot, 1970] considered as system, although the systems methodology is not always used with consistency.

2) Systems approach is at the core of *spatial analysis*, which looked to become the main geographical paradigm in the 1960s [Haigh, 1985]. It did not develop into a general theory, however, and was highly criticized [p. 51] for its abstract approach, essentially concerned with the mechanistic use of computers and losing sight of the human aspect of geography [Olsson, 1979].

3) One of the earliest systems approaches in geography was *geoecology*. This branch of study stems from the original concept of ecology coined by E.Haeckel [Dickinson, 1969], meaning a study of the relations between the organism and its surroundings. The formal expression of ecology in space is *ecosystem*, proposed by Tansley in 1936 [Dickinson, 1969]. The concept of ecosystem has four properties: it is monistic [man, plants, animals, environment in one framework]; it is structured in a rational way; it functions continuously by input and output of matter and energy, and is potentially quantifiable; it has the attributes of a system, tending towards a steady state, together with properties of self-regulation by feedbacks. This approach was at one time considered best suited for the integration of man-nature relationships; geography was even defined as the 'ecology of Man'.

4) *Landscape science* is another systemic trend in geography, which has many advantages [p. 56]. The landscape systemic approach is very close to what I propose as the systemic regional approach.

As early as 1963, Ackerman defined geography's 'overriding problem' as the 'understanding of the vast, interacting *system* comprising all humanity and its natural environment on the surface of the globe.' This concept is very important, because it places the geographer's actual work in wide perspective: the work the geographer is engaged in at a given time is not crucial *so long as the concern is toward a larger goal, however for off it may seem*. Take an ecological study, the dispersion of the tse-tse fly. This, apparently, is an ecological study, but it involves economic realities [breeding of cattle], ways of life [pastoralism], and land-use [agricultural/pastoral]. The study of the dispersion of the tse-tse, in other words, constitutes a study of a component of a regional system, *if the scholar's approach is a systemic one*.

Ackerman summarises the relationship between geography and the systems approach as follows:

a. The universe studied by geographers is the world-wide man- nature environmental system.
b. The world-wide system is composed of a number of sub-systems; these aid
   in identifying a hierarchy of problems for resarch.
c. The techniques of system-analysis aid in applying their space- concept in the
   analysis of subsystems of the world-wide man- environment system.

This trend toward globality is also forwarded by Reynaud [1982]:

> Structure, linkage, interactions, hierarchy are today quite common concepts. It
> is true that geography is long since structuralistic or systemic, at least by
> intention, because geographers, in their desire for globality, claim to study
> correlations.

But geographers, despite dealing only with systems [Brunet, 1979], had seldom
utilised the system approach.

Bennet and Chorley [1978] provided the most integrative attempt to forge a
systems approach to geography [Johnston, 1983], a unified multi-disciplinary
approach to the interfacing of man with nature.

The 'dernier cri' in formalising the systems approach in geography is Wilson's
[1980] definition of five tasks for systems:

1. help in the analysis of complexity;
2. focus on the interdependence of the elements of the system and definition
   of subsystems;
3. mapping of possible methods of analysis;
4. use of the general system theory in modelling;
5. help in planning and problem-solving.

There is a wide support for the systems approach among contemporary geographers. It fits the position of geography as a branch of knowledge with a unique
integrating function [Stoddard, 1965]: geography 'either stands or falls as an
integrative discipline.' According to Huggett [1980], the notions of level of resolu-
tion and hierarchic structure incorporated into the systems approach can provide
the methodology for answering the problem of scale [p. 136].

Langton [1972] writes extensively on the potential and problems of adapting
the systems approach to human geography, coming very close to identifying a
geographical system with a region. A system, in order to be analysed, must be
susceptible to abstraction from reality. But geography deals with *concrete* systems.
A concrete system is distinguishable from unorganized entities in that [Langton,
1972] within the concrete system there should be *physical proximity* of units, a
similarity among units, a '*common fate*' for the units, and a *recognizable pattern* of
units. Those units of reality which interact to produce endproduct are definable
as a system.

The problem of hard and soft systems

One of the problems of adapting the systems approach in human geography is the difficulty of mathematical transformation of the elements of the system. As stated elsewhere [p. 27], some qualities which are nonquantifiable do affect human behavior and are crucial to geographical research. We can overcome this difficulty by conceiving of the systems approach as essentially a *strategy of research*, a *means of structuring complex problem areas* in order to answer or solve identified problems [Morgan, 1981]. The essential task is to *abstract* from the infinite variety of existing systems that one which is most appropriate to a particular problem.

Bennet and Chorley [1978] perceive two basic families of systems, a distinction which can solve the problem of applying the systems approach in social and human sciences [cf. also p. 27]: *Hard systems* are susceptible to rigorous specification, quantification and mathematical prediction in terms of their response; these are systems mostly from the natural sciences. *Soft systems* are cognitive systems, and not treatable by methods of mathematics. They have soft, not rigorous, boundaries, susceptible to integration, and they are not always clearly defined realms. Nevertheless soft systems remain within the definition of a system, as a set of logical operations acting upon [and acted upon by] one or more inputs leading to the production of outputs from the system, by a process of throughput which is capable of either sustaining the operational structure of the system or transforming it.

The concept of soft systems was accepted somewhat grudgingly. It was argued that the soft system is not clearly defined in structure, that the relationship between its components is difficult to assess and that whereas goals, boundaries and procedures are clearly defined and established in the hard system, they are vague and unstable in the soft one. But, on the other hand, the position some held was [Morgan, 1981] that soft systems methodology is more suitable for tackling complex real world problems than the hard systems-engineering type of approach.

The systems approach is a way of solving problems, and geography is considered to be a problem-solving discipline. Among the reasons for the attractiveness of the systems approach to geography are [Agnew, 1984] the hierarchical structure of geographical problems, the lack of closed boundaries between them and their complexity. As the web of interrelationships between man and his environment becomes more complex, the need has grown for a simple framework for structuring and studying the problems. The systems approach can provide this framework as an alternative to modelling all the complexities of the real world.

The first and perhaps most important step in solving an unstructured problem, is [Huggett, 1981] 'to come up with a root definition which captures the basic nature of the system[s] to be relevant to the problem.' This leads to a definition of the method of 'central problem' of a region [p. 91].

There is a difference between formal [hard] and informal [soft] modelling: the formal modelling strategy is rooted in theory, in a paradigm, but in a soft system this is not always the case, as in a soft system repetitive results cannot be always expected: '... in a human activity system repeatable results cannot be expected, owing to *irrationality of human behavior*'. [Huggett, 1981]

Not *all* human behavior is irrational; man can be rational even when dealing with elements steming from irrational realms such as ideology, art, religion, emotions, and so on. Such elements as, for example, 'genres de vie', preferences in food or clothes, family rites, organisation of work, definitively influence the acts of humankind, and this behavior should be taken into account. If it is not possible to formalise these elements into tangible, quantitative expressions, they should be dealt on a verbal, qualitative basis. Irrational elements sometimes have quite realistic expression in the lifesystem: the Mormon religion created settlements and cities in arid Utah, as did the German Templars in the second half of the 19th century and the Zionist ideology in the semiarid and arid zones of Israel. A far more extensive list could be given of the influence of the dynamics of ideologies on landscapes and the processes in them. The apparently qualitative, or 'irrational' elements in human behavior – ideology, religion and mind in general [even economic concepts are based on assumptions of human wishes and perceptions] – have quantitative expression in the landscape. They are a part of the systemic region.

SYSTEMS APPROACH IN REGIONAL GEOGRAPHY

The systems approach is not new in regional geography. An analysis of regional literature reveals that the concept of region was a direct, albeit unwitting, precursor to the concept of systems, even before Bertalanffy's formalisation of the systems concept. The trend towards integration of elements from different realms has, in general, characterised regional geography. The systemic region ['la région sys-temée'] represents the Vidalian idea of interaction:

> The assimilation of a region to a system, or to a product of a system, was in geography a latent model, anticipating in its fundamental intentions the general theory of systems [Dumolard, 1975].

Nevertheless, the systems approach was not applied to any great extent in regional geography [Beaujeau-Garnier, 1972]. There was no structuring of a systemic model, despite the many favourable opinions towards the systems approach.

The systemic approach is a 'gymnastics of the mind' ['gymnastique de l'esprit', Reynaud, 1982]. It is the only tool which can explain and solve contradictions between the nomothetic and the idiographic, the general and the particular, the totality of the globe and the local individuum. Elsewhere was discussed [p. 23] the theoretical basis attributing to the region the nature of a system, considering it as a 'holon'. I also tried to show that the systems approach can answer the basic geographical dichotomies [p. 5]. The system, being structured on sub-systems, can also solve two other geographical dilemmas, those of scale and of the relevance of variables to be included in a study.

The problem of scale

The problem of scale is the choice between extension of or concentration on the object of study [Watson, 1978]. The question of what scale is optimal for a certain study has been answered mostly arbitrarily and not always in harmony with the nature of the research. By adopting the systems approach, the scale will accord with the size and nature of the system. The possibility of discerning between a system and its subsystems is a precious tool for adapting the scale to a certain problem. If the aim of the research is a study of a system without entering into the particulars of its sub-systems, then it is possible to choose a small, generalizing scale; if the aim is to enter into a sub-system, then a larger, more detailed scale can be used. The very substantial dilemma of going into depth or width can be solved on a rational basis by this approach because, by definition, one is or is not undertaking a study of sub-systems of the subject-matter. In a rather subconscious way this has been done by some of the regional studies which dealt with larger units: first they dealt with the general characteristics of the region [system] with the scale appropriate to it, and afterwards with the sub-regions [sub-systems] on a more detailed scale.

Take a study of a city. One of the many chapters of such a study can be its road system. Let's say that the scholar thinks there is no need to deal with the road sub-system in more detail than other urban sub-systems, and decides not to enter into the technological, economic or operative details of it. But if the need arises for a more detailed treating of the road system – because it was disclosed to be of special importance for the functioning of the whole urban system – it is possible to enlarge the scale of the study of this particular subsystem in comparison with other sub-systems which are not significant to that degree and therefore do not require a more detailed study. In another city, perhaps another urban sub-system – water supply system, educational system, employment – will be more crucial than the road system. The systems approach gives us the epistemological argument and the methodological tool to fit the scale to the nature of the study.

Relevance

One of the difficulties of regional geographic research – and, to be frank, one of its handicaps which was, sometimes with justification, heavily criticised – was the trend to say everything about anything. This misuse of regional geography introduced as many items as possible, even those which had no relevance to the study, creating, in fact, an inventory of elements only because they were present in the area. This approach should be eliminated from a serious regional study, even if today people – mostly non-geographers – do write from such a perspective.

Only elements relevant to the study should be treated; the irrelevant should be omitted. The problem has always been that of discerning between relevant and irrelevant elements, between the essential and the incidental. The decision, obviously, has been very subjective.

The systems approach is a definitive answer to the problem of relevance. By definition, a system is not an assemblage of items, but a set of components tied together by the relationships expressed by the processes acting in the system. The difference between an item and a component is that a component, active in process, *is a necessary part of the system*, whereas an inactive item is not a necessary part of it even if it does exist in the area. This approach allows us to decide, in a non-arbitrary fashion, whether to include or exclude a certain element in our research. Sometimes this decision can be aided by a pilot-study. Take, for example, a certain mineral deposit, such as coal, iron, underground water, etc. When embraced within the economy of a region it certainly is a relevant part of it, not inherently, but due to man's use of it. As only one of the components of the system, it is a part of it, if and when it contributes to the regional processes.

Mineral deposits lying deep under the surface and not used by man have no relevance to the region and to its functions [Gottmann, 1957]. Perhaps in the past such a relevance existed [abandoned gold fields in New Zealand, Alaska, California] or will be attained in the future; but if at the time of the study the deposit is not being used, it is irrelevant to the region at that moment and should be neglected in the regional research. Each relevance is linked to a certain historical time; it is impossible to disconnect the study of space from the study of time.

Perception of holons and systems: vis-à-vis 'Noce de Cana'

To understand the nature of the term 'system', it might be helpful methodologically to compare the way in which we, as viewers, regard a painting in museum to the way in which the painter sees it [Weiss, 1969]. To see the picture it portrays, it is necessary to have a certain distance from the painting, allowing one to see the different parts converging into a whole. On the other hand, to see the different parts of it in detail, one must come quite close to it. To perceive both the whole and its particulars, the apparently absurd requirement is that one be at close to it and at a distance from it.

When we approach very close to a picture, as close as we might hold a book in front of us to read, all we see is different spots of colour, sometimes points of color only millimeters in size. These are the smallest parts of the matrix, let us say, the 'holons' of it. Each of these points of colour is a component of a greater holon which, in turn, is a part of a still greater one. All the points, all the holons, become a part of a whole when one's distance from the particular holons increases and one's gaze can embrace a larger space, meaning a greater number of holons. Perceiving a particular dab of colour within the picture is essential to the perception of the whole picture; but if we are unable to perceive the connection between the individual dab and its neighbours, with other holons, we cannot perceive the whole picture. It seems quite natural to consider a picture as a system, more than simply the sum of its points of colour, its holons. A picture creates a certain relationship between the different points of colour. It is this relationship between the components and the whole which makes a particular painting different from others, even those created through the same medium.

When we look at a large painted scene – let us say Veronese's 'Noce de Cana', which is perhaps one of the largest paintings in the Louvre [Fig. 5] – we can only

Fig. 5. Veronese's 'Noce de Cana' – Louvre.

perceive it as a large system in which each holon, each point of colour, is integrated. More than a hundred and twenty persons and items figure in this painting, and each of them is a part of a system, as perceived through the artist's mind. The picture would be quite another system if some of the persons or items now depicted did not figure in it. In the mind of the artist, each component had a distinct function in this system: the dwarf in the foreground, the campanella in the background, the massive but elegant pillars, the richness of costumes and meals, all are parts of the system. Many other paintings depict this theme of 'Noce de Cana', but in them the aim of the artists, the 'system', is different, and therefore the components are different.

As stated previously, the observer must have a certain distance from the painting to be able to perceive it as a whole, and have a certain closeness to it to perceive it in detail. To analyse the details, he must be so close that he can perceive only the particular holons. But when analysing the holons, he knows that the holons are but parts of a wholeness, of the whole picture, and this knowledge is an integral part of his perception and analysis. The spectator can also concentrate on an analysis of the different components of the picture: an analysis of the costumes, or of the different architectural elements, or even more theoretical themes such as how the light is translated by the artist, and so on.

If we compare the idea of a systemic region to a complex painting, it seems to provide us all the didactic results we expected. Each element of a region, being in relationship to other elements and participating with them in the whole system-region, is like the point of colour which is the smallest element of the painting. But as the spot has meaning *in this picture only as a part of the whole*, so the holon has meaning in a region only in its relationships with other holons of the region. And, as in a painting we can analyse different holons by different criteria – part of the whole, part of subsystem, part of a theme distributed on the picture – so different holons can be treated in their geographical distribution.

The dual approach to a picture, from near and from afar, is a must for the artist, but is also of great benefit to the observer who wishes to understand a picture in its deeper meaning. An artist is blessed with the great gift of seeing the particular, sometimes miniscule spot of colour not only as itself, as a small holon, but with anticipation of its meaning when surrounded by other holons. As he paints the particular holon, the artist has to be able to see it also as a part of the whole, which then exists only in his mind.

The great achievement of the systems approach in science, in social studies and in humanities, is to see the particular as a part of a whole, and to see the whole as composed of parts. This way of thinking can form a bridge between the arts and science. Indeed, it is a way of expressing the existence of the wholeness of which human existence is only a part.

Is a picture a closed or an open system? The answer seems obvious, as of course a picture is framed, sometimes by a frame which makes the edges of the picture one of its most rigid elements. According to what we see, the picture is a closed system. But let us look, for a moment, beyond the tangible elements of a picture. Is it only the formalised use of colours which makes a picture? Is it not also the effort of the artist to give a visual expression to certain concepts? Of course a picture is a system closed spatially in two dimensions, but it is not closed in depth and time. A picture contains a multitude of feelings and beliefs, based on education, experience, intellectual and social background. All these concepts exist behind the picture. Therefore, only apparently is a picture is a closed system, for *it radiates influence outside the frame* towards the mind of the observer, and, moreover, is a *product of ideas stemming from the artist's mind*. So it is with the systemic region as well: although a region has a frame, a limit, a boundary, which seems apparently rigid, a regions can only be an open system, the type and magnitude of its openness depending on many factors.

To continue a geographer's visit to the Louvre, we might linger a while in front of Louis Le Nain's painting 'Peasants' Family' [Fig. 6]. The systems approach should help us understand the intentions of the Master and perceive the radiation of its influence *outside the frame*.

At base it appears to be a scene of desolation, or at least of profound acceptance of fate. We see the deeply introverted contemplation of the father cutting the bread; the no less desolate, but protesting look in the eyes of the mother; only the child has naive, wide open eyes. The atmosphere is grey, the colour grey in different nuances pervades the painting. The only vivid colours are the red wine in the glass, and a little spot of green on the shirt of the young woman. All the elements, the holons, of this scene serve the system, which is a *depiction of misery endured with dignity*: one both sympathises with and admires this family. To achieve this impression, a great artistic mind using consummate skill created the smallest

Fig. 6. Louis le Nain's 'Peasants' Family' – Louvre.

material holons, points of colour, by perceiving them as they would appear in a whole, as parts of a pictorial system, to elicit a mental reality, sympathy and admiration for a peasant family of the seventeenth century.

TOWARDS A MODEL OF THE REGION AS AN OPEN SYSTEM

The problems are solved not by giving new information, but by arranging what we have always known.

L. Wittgestein

THE MODEL OF RITTER AND HETTNER

Since the dawn of classical geography, but particularly since the 'age of discoveries' in the fifteenth century, geography has fulfilled the primary, basic task of scientific procedure: gathering of data, recording them and classifying them. Geography needed to document and describe new territories as they were discovered, in the way botanists and zoologists in a parallel situation were occupied with systematic description of plants and animals. During the nineteenth century regional geographical thought was crystalized by Ritter [p. 34] into a certain way of thinking which became the model of regional reasoning and lasted, after being further formalized by Hettner [p. 37] until the beginning of the twentieth century.

This model was based on the assumption that geography studies the influences on the life of man of physical elements such as climate, topography, distances, etc. It also fit the then supposed link between geography and history; geography was viewed as having the task to explain, within the framework of a certain environment, the physical background to historical events. Therefore this model [or 'Laenderkundliches Schema'] dealt first of all with the explanation of the boundaries of a certain country or region, the next step being the explanation of the physical background: geology, lithology, pedology, climatic conditions, hydrology and topography. All these phenomena were considered the basis for the distribution and activity of population. As most of the world economy during the nineteenth century was still based on agriculture, industrial societies comprising only a small fraction of the world's population, it seemed logical, natural and obvious that the physical background is what decides the character of a region. Moreover, even industry, which in that century was concentrated in proximity to bulk mineral resources, such as coal and iron, which contributed to 'heavy industry', it seemed reasonable to consider the physical environment, including the 'natural resources', to be the decisive factor in the distribution of anthropic phenomena. The anthropogeographical phenomena – population, government, religion, etc. – followed later.

This model was not only a methodology but a way of thinking, as Hettner formalized it [Hettner,1932; translated by D.N]:

> To begin with, it is clear that the 'Scheme'is nothing other than a reproduction of the natural realms: the three realms of anorganic Nature, of Plants and Animals, and of Humanity, and inside each realm a reproduction of the manner of their appearances; so, for example, regarding the solid surface of the earth, it is the form, the substantial qualities and the processes; in the realm of plants, it is the vegetation and the flora; regarding man, it can be settlement and villages, transportation, economic activity, way of life, races and nations, religions and states. The common sequence of the natural realms also conforms to Nature itself in that is impossible to imagine or to conceive of the existence of man without Nature, the realm of the animals without the realm of the plants, and either of them those without anorganic Nature, the air and water without the solid surface of the earth. They are together as floors of a building, to which a land can be compared.'

> [Zunaechst is klar, dass das 'Schema' nichts anderes ist als eine Wiedergabe der Naturreiche: die drei Reiche der anorganishen Natur, der Pflanzen- und Tierwelt und der Menscheit, und innerhalb jedes Reiches eine Wiedergabe der Erscheinungsweise: z.b. bei der festen Erdoberflaeche der Form, der stofflichen Beschaffenheit und der Vorgaenge, bei der Pflanzenwelt der Vegetation und der Flora, beim Menschen etwa der Besiedelung und der Ansiedelungen, des Verkehrs, des Wirtschaftslebens, der Lebensfuehrung, der Rassen und Voelker, der Religionen und der Staaten. Auch die gewoehliche Reihenfolge der Natur-reiche entspricht insofern der Natur selbst, als der Mensch nicht ohne die Natur, die Tierwelt nicht ohne die Pflanzenwelt,diese nicht ohne die anorga-nische Natur, Luft und Gewaesser nicht ohne die feste Erdoberflaeche moeglich und infolgedessen, wie ich es frueher ausgedrueckt habe, vorstellbar oder denk-bar sind. Sie sind gleichsam Stockwerke des Hauses, als das wir ein Land auffassen koennen.]

This 'scheme' was challenged by Spethmann in 1927, 1928, and 1931. The titles of Spethmann's books, 'Dynamische Laenderkunde' [Dynamic Geography, 1928] and 'The Geographical Scheme in Germany: Struggles for Progress and Freedom' [1931] express Spethmann's attitude towards radically changing the 'scheme'. His proposal was to use this well established model *to analyze the forces* which act in the region, beginning with the most important ones, *most of these latter generated by man*. His books and proposals were heavily criticized by Gradman [1929] and by Hettner [1932], as the German geographical establishment of that time, personified in Hettner, did not accept an undermining of the ruling model. This continued to represent the main regional model until the 1950s, when it became the principal target of the criticism of regional geography [p. 44].

THE 'CENTRAL PROBLEM' MODEL

This model was more preached than practiced, but still it contributed to regional geography some of its most important contributions in the 1940s and 1950s. It comes from a more or less didactic concept: out of the multitude of phenomena existing in a region, the representative, decisive, basic one should be the center of study, the focus of the scholar's efforts, the subject of analysis and verification [p. 125]. This phenomenon is the central problem of the region, and influences the totality of the area. This approach is quite close to Spethmann's postulate of dealing, in the first instance, with the forces ['Kraefte'] acting in the region.

Gourou focused his attention on the peasantry of Tonkin in the 1930s [1936], and on land use in the tropical world [1947/66]. Gottmann bent his efforts toward Virginia's cultural heritage [1957] and the urban network of the northeast United States [Megalopolis, 1961]. These are perhaps the most important examples of the application of this model.

Focusing attention on a central problem as a methodical is advantageous in that it facilitates a sorting of the different variables in a region according to their apparent connection to the central problem. In this way, it enables us to clear away much of the useless 'noise' which heavily burdened the classical Ritter-Hettner model. The logical extension of the central problem model is the model of systemic region.

THE MODEL OF SYSTEMIC REGION

In this model, I wish to amalgamate the thoughts expressed above into a model capable of moving regional study beyond the traditional approaches applicable to different types of regions. Believing that most systems are structures of sub-systems, I do not intend to propose a world-wide system or one which includes large sub-systems. Even a state – excepting perhaps the smallest ones – is a large system of sub-systems. The basic idea of this model is influenced by Wilson's model on the urban system [Wilson, 1981, p. 265]. The proposed model seems to fit a study of a rather modest basic unit – a county, a district, or a part of them – as well as a town or city; these can be considered systems of sub-systems [Nir, 1987]. The three parts of the model [Fig. 7] are input, regional structure and output.

Input

To day no socio-geographical system exists which could be defined as a closed one, detached from its environs. Perhaps some closed societies can be found in the remote parts of Amazonia, but such structures should be considered relics. There can be, of course, a voluntary separation from the outer world for political, religious, ideological or other reasons, but even this detachment cannot be total [Albania, Iran]. Therefore, regions must be considered as open systems, as they

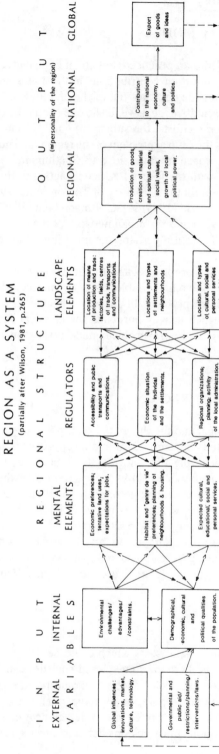

Fig. 7. The model of the systemic region.

are – even in cases seeking with great care to control their identity and character [Nir, 1985] – influenced by the outside world, this influence being part of the input of the region.

In the literature, a distinction is drawn between internal and external input [Chorley and Kennedy, 1971], the internal originating in the system itself, the external coming from the outside. There are two possibilities of external input. One type stems from a permanent, regular source supplied to the system by its geographical and political situation, which we call *national input*. The second consists of irregular, sometimes uncontrolled, distant sources that we call *global input*. The national input is the sum of influences affecting the region as a part, a subsystem, of a state or a nation: interventions by national or public institutions, legislation, national planning, development projects, governmental aid or limitations, flow of capital and information. Global input is information, technology, capital, initiative and culture of other than national origin that enter the region. This input is not expected to be as regular as the national input. Sometimes its flow is hindered by national, political or economic constraints. Global input also includes the demands or limitations of the world market, technological, cultural and scientific innovations, investments, exchange of knowledge in science and education, foreign aid, and so on. No region exists today which is altogether lacking in global input. Important decisions affecting the life of a region – a decision to build an industrial plant, for example, or to establish a national highway – can be decided outside the region, in a very distant place perhaps, and sometimes without any initiative from the local population. External inputs, although having a spatial origin outside the topographical boundaries of a systemic region, are an important part of it which must not be neglected.

The *internal input* is composed of both the environmental and the human – demographic, cultural, economic and political – qualities of the region. These, of course, are influenced by the national and global inputs, as there exists a flow from the external to the internal input, albeit in some cases controlled or hindered. *We call the internal input the challenge of the region.*

Environmental challenges can stimulate or constrain the activity of the population and are integral parts of the regional input. This consideration makes our model one of regional geography, and not of regional science which, as explained elsewhere [p. 55], does not deal with environment at all. Environmental challenges exist, even today and even in highly developed countries, but the answers to them have become routine, and we pay no special attention to them except when we are reminded of them in critical moments. A mutual feedback exists between population and environmental challenges: an environmental element considered in the past to be a geographical constraint today becomes an economic stimulant and advantage, or vice versa. The perception of snow in the Alps seems to be a good example of this [Isnard, 1985]: only a few decades ago, snow was a negative element in the Alps, limiting the movements of population, transportation, economic activity, and other aspects of human existence. Today, the snow is an asset in certain alpine regions, being a necessary element of the ski industry. Its abundance is considered a blessing. The change in attitude towards snow – a feedback in the man/environment relationship – made possible the revival of some montainous regions which had been on the fringe of demographic degeneration.

The regional structure

The center of the model is the regional structure, composed of *mental elements* [Wilson, 1981] of the region which by the intermediation of *regulators* become the visible *landscape elements*. When we enter a region or deal with data on a region, the first elements we perceive visually or mentally are the *landscape elements*, which are the objects on the region's surface: buildings, vegetation, people, roads, topography, etc. For generations, these tangible elements were the object of geographical description. But the aim of geographical explanation is to also see the mental elements behind the objects. Although intangible, they are, nevertheless, basic to the landscape: motivations, ideologies, national and private tendencies and aims, which are formalized and realised into what we see as the landscape elements. A true geographical study must go behind the immediately visible and study the degree to which the mental elements are realised or modified by the regulators.

Following Wilson [1981] *mental elements* are placed into three groups: expectations of employment and production, i.e., economic preferences based on certain personal or group ideologies; preferences of habitat and neighborhood; and preferences for services, education and quality of life. Ideological or iconographical structures are a vital part of the mental elements.

The perceptions, expectations and economic preferences are challenged by the *regulators*, which can be positive or negative, stimulating or constraining: accessibility, topography, climate, as parts of the environment; economic capacity, cultural assets and social situation of the individuum or of the municipal or social unit of which one is part; local and regional level of organization, planning and activity. *The landscape elements, as we see them in the region, are the expectations and preferences of the present, and even past, population, as metamorphised by the regulators.* The regulators are both inherent to the system and also influenced by elements of the external input, such as governmental and global influences on the activity of the local population.

The landscape elements are, as stated above, the tangible, visible objects in the landscape. They are the results of the processes between the mental elements and the regulators: the localities of work and employment such as factories, fields, shopping areas and commercial centers; means of transport such as roads, railways and airports; residential quarters including houses, parks and gardens, educational, health and recreation centers.

The regional structure seems to be an essential segment of the model, being the visible scene of processes and interactions. The flow of information and of means is necessary to maintain the processes acting between the expectations and the landscape elements. Together with the feedback – learning a lesson and putting it anew into the flow – which is an inseparable part of the processes, the flow of information and means is the main artery for the linkage between single components.

All the elements characterising the function of a system, as explained in the literature [Chapman, 1974; Langton, 1972; Wilson, 1981], exist in the region. Dynamic regions exist in which the processes change the region's basic features; their character remains the same in some regards, but in other regards is different from their previous nature. Some regions have a certain stability of landscape elements, where a 'steady state' prevails as the processes and feedbacks preserve the constant nature of the output. There are regions ruled by entropy, where only minimal native processes preserve only a minimal flow. And there are dying regions where every activity is in decline, and the main process is a deterioration of the population and of the landscape elements. The systems approach proves its validity on each level.

Output

According to the model, there are two possibilies of regional output: output which remains in the region, and output having impact beyond the boundaries of the region in which it was produced. What has been said of the external input can also be said of the external output: it may be national output, affecting the nation, or global output, which is to say that potentially the output of a certain region is being exported on a global scale.

The nature and value of a region's the output depend less on the input than on the regional structure, its components and processes. Perhaps herein lies the importance of the study of regional geography: to consider *what has been done by a population possessing certain mental elements, exposed to external input in an environment*. It can happen that in a certain region great efforts and means are invested by external input, with, however, disappointing outputs. The opposite can also occur, that the efforts of the population in a region lead to an output beyond all expectation according to the initial input.

A region may have output which is poor and meagre, most of it remaining in the region itself; on the other hand, the output of a region can flow outside of it, to the nation and around the globe. *By 'output' we mean all the creativity and productivity of a region*, be it tangible goods, such as merchandise, raw materials and food, or spiritual values and creations, including culture, political or social ideas: it should not be measured only in economic achievements. A nation may invest in a region because of a special interest in a particular output of the region, as, for example, investment in a border region where the output is strategic rather than necessarily economic. A nation may aid an economically feeble region, so that its will contribute to the social homogeneity and political stability of the nation.

Classical vidalian geography designed the characteristics of a region by the term 'personality of the region'. It seems to us that the output, which is the result of the processes acting in the region, can be considered as a part of such a 'personality of the region.'

Conclusion

I have tried here to apply the systems' approach to a model of systemic region. This approach bridges the physical and socio-anthropic components of the region; it can use methodologies appropriate to both of them; it makes it possible to deal with regions large and small, as it accepts the existence of hierarchies. This approach aids in the identification and study of the dynamics of regions, discerning between active, stagnant or declining ones. It facilitates the understanding that *change is the most stable process*. This approach is a valuable springboard for planning, for political fact, for intellectual curiosity, because regionalisation, which is the pragmatic aspect of regional geography [p. 156] can point to a certain region as different from its neighbours and thus as an object for development and improvement. The social efficacy of regional geography is in promoting more equality among regions [p. 158].

APPLICATION OF THE MODEL: THE REGION OF BET SHEAN, ISRAEL

I have applied this model of systemic region to the region of Bet Shean, which I have been studying since 1956 [Nir, 1968; new edition in press]. The region of Bet Shean is a municipal authority, the Regional Council of the Valley of Bet Shean [Fig. 8], an administrative unit consisting of twenty-one villages [sixteen kibutzim and five moshavim]. In the middle of this area, actually its geographical as well as its population center of gravity, is situated the town of Bet Shean, which constitutes a municipal entity in itself. In this semi-arid environment, we find intensive modern agriculture on 14,000 ha, nearly 90% of it irrigated. The agricultural products are treated in the plants of the regional enterprises, where most of the products are packed ready for market. In 1982 the value of the products shipped by the regional enterprises reached $100 million. The town of Bet Shean developed from a small Turkish village of the beginning of this century into an organized township. There are industrial plants both in the town and in the kibutzim, in some of which industry accounts for more than 50% of the income. According to our model [Fig. 9], in order to understand these tangible landscape elements, we have to study the mental elements behind them, the regulators, and, of course, the input of the region.

The internal environmental variables of the *input* are a semi-arid climate – very hot in summer and mild and rainy [240–320 mm rain annually] in winter, chalky soil, and a hydrological asset in the form of a group of springs which yield nearly 90 million cubic metres of water per year. These springs are the most important single environmental variable which affects the whole existence of the region, making possible intensive irrigation. The internal demographic variables are the ten thousand inhabitants in the rural and fourteen thousand inhabitants in the urban sectors. To understand the forthcoming section on mental elements, we must distinguish, as a part of the input, between the collectivistic ideology which

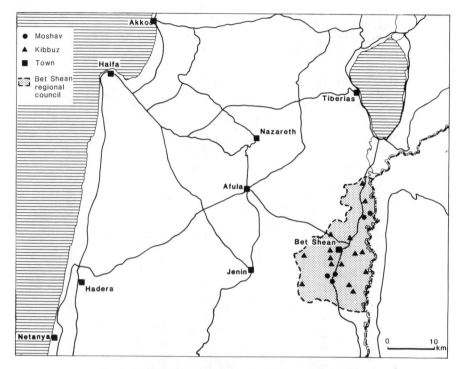

Fig. 8. The region of Bet Shean, Israel: map of orientation.

is part of the human input in the kibutzim, and the private motivations which, of course, inhere in the iconography of the inhabitants of the town. A mixture of individualistic and cooperative approaches characterizes the moshavim, or the small-holders settlements.

Among elements of the external national variables we can cite, as examples, the national boards of citrus, vegetable, and cotton production, the nationalisation of the water resources, the official policy on the cost of water, etc. The external global variables are more difficult to trace, but, the price of cotton in the world market is one example among many of a global variable which influences the production of cotton in the Bet Shean region, where cultivation of cotton occupies nearly 30% of the arable land. The drop in cotton prices in 1985 induced a reduction of the area of cotton cultivation in the Bet Shean region from 5000 ha to 3000 ha.

The *mental elements*, as mentioned above, differ markedly between the urban and rural sectors. In fact, the model revealed that this systemic region is built of two sub-systems, one urban and the other rural. To be more precise, the collectivistic and the individual 'Weltanschaungen' create two 'genres de vie' in the same region. In confrontations between the mental elements and regulators, each

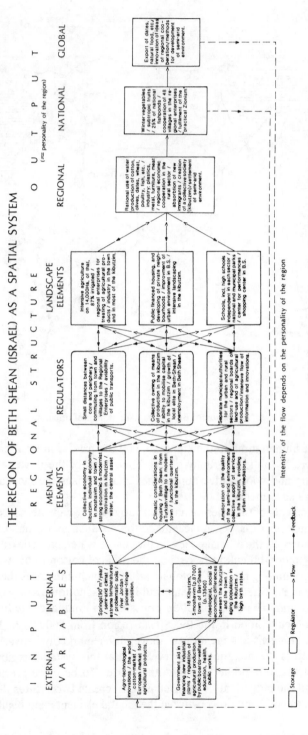

Fig. 9. The model of systemic region applied to the region of Bet Shean.

sector responds differently. The mental elements of the rural, predominantly collectivistic sector include ideas of modern agricultural techniques, cooperative processing and marketing, combining agriculture with industry – following Kropotkin's ideas [Kropotkin, 1902], which are dear to the kibutz ideology [Nir, 1986a] – and high expectations of cultural and social services. Most of these ideologies are strange to the urban sector of the region.

The mental elements of these two sub-systems of population are confronted with the regional or local regulators. The realization of the mental elements depends on topographical regulators, such as the rather small distance between the villages and the regional center. A very important regulator in the rural sector is the regional organization of the villages. In the urban sector, the most important regulator is the governmental initiative in bringing in industry and social services.

To illustrate the interaction between the input, mental elements and regulators, consider the utilization of the water sources, 40% of which are rather saline and require special treatment. The *input* consists of 25 sources of different levels of discharge and quality in the northwestern corner of the region. The *mental elements* include the consensus to accord to each farmer an 'equal portion of water of equal quality,' and the responsibility of the region's inhabitants for the management of the sources. The *regulators* are governmental aid in establishing a 'mixing plant' where water of different qualities is mixed into a quality acceptable for irrigation. The *landscape elements* resulting from these interactions are a network of concrete canals, pipes and water reservoirs criss-crossing the region, including the water supply to the town, and, of course, the output is the capability to irrigate 90% of the arable land. In fact, the entire structure of the agricultural economy is based on this regional organization of the water supply.

As defined in our model, the output of a region is all that is created in it. The regional, or local output is, first of all, the way of life within it and its products. In the Bet Shean region, the agricultural products are fish, dates, winter vegetables, cotton, grapefruit, poultry, 'natural' food, etc; the industrial products include plastics and electronic equipment. Its output is also the regional organization of the villages, which here attained a high degree of cooperation in regional enter- prises, an accomplishment which serves villages even outside the region [Fig. 10]. The output is certainly not restricted to the region; most of the goods produced are shipped to other regions of Israel and overseas. But besides marketable goods, its less tangible output is equally, perhaps even more, important: innovative agrotechnical methods, especially in irrigation; the cummulative experience in the settlement of semi-arid zones; the efforts in acclimatization of plants; and, last but not least, its experience in regional cooperation.

A closer examination of the components of the region – as illustrated by analysis of the mental elements and shown in the different stocks of the model – leads to the conclusion that in this systemic region there exist two sub-systems. Within the current social situation in Israel, this model indicates that the 'central problem' [p. 91] of the future may be the development of the relationship between the rural and urban segments. The 'central problem' of the mid-fifties, which we defined [Nir, 1968] as the rational management of the sources, today has become routine.

100

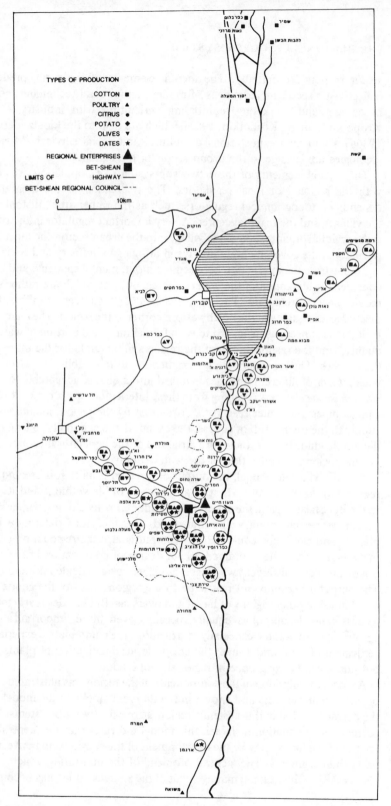

Fig. 10. The regional organization of processing of agricultural products: villages incorporated in the regional enterprises of Bet Shean.

The relations between the two sectors, rural and urban, today are rather restricted. The kibutzim, because of their collectivistic structure, do not buy and trade with the relatively small town. Instead they deal with national networks of production and supply of which they are a part. Fifteen percent of the active population of the town work in the regional enterprises and this is the only socio-economic link between the two sectors. In the future the relationship between them may intensify through stronger ties in the cultural, social and economic realms; on the other hand, political differences, which are not inconsiderable, and reflect disparate opinions vis-a-vis the collectivistic way of life, may enlarge the gap between the two sectors.

The perception of one regional system structured in two sub-systems can enlighten public opinion on the possibilities for further development, common services [for example, environmental quality], and cultural and economic cooperation, as in the development of tourism which affects both sectors. It does occur that within a system an *inclusion* [Miller, 1978] develops, which means that a part of a system becomes detached from it, despite existing within its physical body. Such a development is more than undesirable when harmonious coexistence of a pluralistic society is the aim of public policy.

Seeing the different elements of a region not as detached units, but as components of a system – in our case-study, a system consisting of two definitively structured sub-systems – makes all the standpoints related to the region more involved in solving regional problems.

Study of a topical theme within the model of systemic region

I would like now to demonstrate how this model can overcome the apparent dichotomy between regional and topical geography. A region being a system composed of sub-systems, the scholar who studies a region can concentrate his regional interest on one of the sub-systems encompassed in the region, *keeping in mind that this sub-system is a part of a regional system.* By introverting or extroverting the study, it is possible to bridge the regional with the topical subjects.

A case study of this approach is a study of industry in the kibutzim of the Bet Shean region [Fig. 11]. As part of a regional study, our concern with the industry in the kibutzim [rural settlements] has to do with its role in the socio-economic structure of each settlement and in the region: its part in the income, its infrastructure, its influence on both the social relationships within the kibutz and the nature of these formerly totally agricultural settlements. All these questions are relevant to the regional sub-system of industry. But the industry in the kibutzim is also a branch of a spatial system or systems: the industry in the entire kibutz movement, as well as the entirety of Israeli industry. When our interest is in one of these systems, the questions and subjects of our research will, of course, be different.

If our approach is a systemic one, each component of the subject matter can be tackled as being a part of both a regional and a spatial system. We can ask what part the industry of plastics in the region of Bet Shean plays in the plastic industry

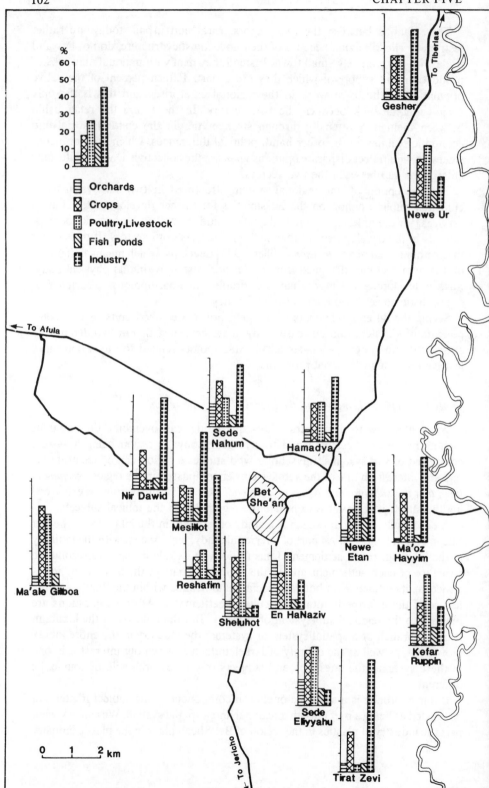

Fig. 11. Industry as a part of income in the kibbutzim.

of the kibutz movement, in the plastic industry of Israel, and so forth. The regional study is a necessary base for such an inquiry. Moreover, the two systems, the regional and the topical, *meet at a certain concrete physical place* where our object of study is anchored. If we are considering this place only for itself, without its ties to space, there is no possibility of systemic spatial treatment; on the other hand, dealing only with the space without admitting that it is composed of places leads the study to the same sterile ends as does the study of place without linking it to space.

The model of systemic region as a way to study the 'hidden factors'

A closer analysis of Fig. 11 reveals that although in most of the kibutzim industry occupies an important share of the income, as much as 50 and 60% of it, nevertheless some kibutzim have no or practically no industry. In order not only to demonstrate but also to explain facts, we must try to understand, for example, the great difference in this regard between two neighbouring kibutzim, Tirat Zvi and Sdeh Eliahu, in the southeast of the region. In the former industry accounts for 66% of the income, in the latter only 9%.

The first step of a positivist, fact-finding inquiry is to check the concrete phenomena: the physical background of the agriculture, such as climate, soil, water resources. In both kibutzim these components are almost identical, and the preference for industry in the one cannot be attributed to agrotechnical failures.

The next step would be an analysis of the socio-economic background. In this regard, the two kibutzim also have common variables: they were established in the same period [1937 and 1939], by the same type of settlers [Jewish immigrants from Germany] and financed by the same institution [The Jewish Agency for Palestine], and they have the same cultural and political ideology [both are members of the religious branch of the kibutz movement, linked more or less to the same political party].

This inquiry does answer why industry is predominant in Tirat Zvi and nearly absent from Sdeh Eliahu. The answer could only be found after delving into the personal opinions and 'iconographies' of the inhabitants, which requires local investigation; the statistical data provides no clue. In Sdeh Eliahu the majority of the inhabitants believe that Zionism – the return to Zion – means return to the soil, to an intimate relationship with it – agriculture. This opinion, with no consideration of economic consequences, is the reason for not introducing industry into the kibutz. As a matter of fact, the kibutz is thriving on its primarily agricultural base.

The use of this model, which forces us to investigate, in addition to the tangible phenomena [= landscape elements], the mental elements as well, can bring us closer to the truth and to a relevant explanation. The 'hidden factors', which are eliminated by the method of 'great numbers' and disregarded as 'noise', can provide the most accurate answer.

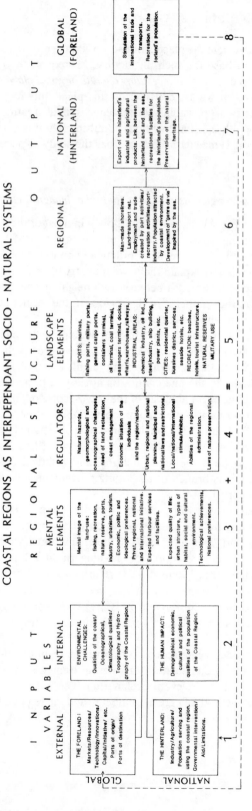

Fig. 12. The model of systemic region applied to coastal regions.

APPLICATION OF THE REGION MODEL TO A SINGLE-FEATURE REGION

A single-feature region or a zone [p. 63] can also be treated succesfully on this model. I applied it to a general study of *coastal regions* [Fig. 12]. The coastal zone is exposed to challenges from the interaction between natural elements and anthropic alterations of them. Any change induced by man in the coastline generates a reaction in the functioning of the natural processes. I wish to stress the need for perceiving this interaction if we are to live in coastal regions with minimal harm to both the natural environment and the socio-economic structures existing there.

The concept of internal and external input is quite apt for studies of this type of region [Nir, 1986b]. The internal input includes the environmental challenges of the coastal region, the quality of the coastline, its oceanographical qualities, and its population. The national and global input [and output] correspond to the concepts of hinterland and foreland, used in the geography of coastal regions and ports. The coastal region is not an isolated, closed system within its given limits. The hinterland is the continental economic and political continuation of it, and the foreland is the realm of its connections, linkages and activities beyond the sea. The hinterland is in most cases the nation to which the coastal region belongs, whereas the foreland is, practically speaking, the surface of the globe with which the coastal region has relationships.

There are mutual relationships between the coastal region and its hinterland and foreland. When the hinterland is developed and needs an outlet for its products, important ports are created. If the coastal region offers sandy beaches and the hinterland has a population appreciating this asset, coastal resorts and recreation centers can develop. Change in the function of one of the elements composing the internal or the external input affects, in a certain manner, the landscape elements of the coastal region. A change in port technology introduced in a distant port, the opening of a new market or a new industrial plant in a distant foreland, a change in the land use of the hinterland, all can induce changes in the landscape elements of the coastal region.

Mental elements of the population of the coastal region have to do with the conception of the purposes of the land use: economic and political preferences for establishment of industry and ports, or for creating natural reserves, centers of recreation and tourism. Natural hazards and oceanographical and morphological constraints, the economic situation of the population, and so on, constitute the regulators. The landscape elements of a coastal region are very characteristic: commercial and fishing ports, marinas, military ports, general cargo ports, containers' terminal, as well nature reserves, military installations, seaside hotels, etc. – all results of the confrontation between the mental elements – fed by the input – and the regulators.

Regional output consists of changing the coastline by human intervention, employment created by port activities, recreational activities, commercial fishing. National and global output, as explained above, is very far reaching.

This case study only hints at the many possibilities of applying the model in different aspects of regional geography.

THE PHILOSOPHY BEYOND THE REGIONAL MODEL

Philosophy – the highest aspiration of the human spirit

Lauer, 1965

Aucune idéologie n'est scientifique, mais toute science a un contenu idéologique.

Nicolas, 1984

GEOGRAPHY, POSITIVISM AND PHENOMENOLOGY

Under the term 'philosophy' we comprehend the study of relations of beings to their existence. There exists a philosophy of the individuum's relationship to the purpose and meaning of his life; there exists a philosophy of ethnic, social and cultural human groups, which relates to their meaning and aims; as a result no a field of human action exists that does not have philosophy.

An individual's philosophy is [Harvey and Holly, 1981 b] both a personal attitude towards life and the universe and a method of reflexive thinking and logical inquiry in the attempt to develop a view about the whole system of existence. As the individual's belief system is central to his philosophical viewpoints, there is a diversity of philosophies:

> As long as there are people with different beliefs, temperaments, and as long as they have something to say, there will be place for different schools in philosophy [Couclelis, 1982].

Philosophy attempts to gain a comprehensive view of things. Whereas science is more analytic, philosophy is more synthetic, bringing things together to an interpretative synthesis [McDonald, 1966], and aims to discover the significance of things. Philosophy means a search after a meaning which is beyond every day use and behavior. It aims to explain the every day in a certain perspective which sees more than the apparent phenomena [Bunge, 1976]. As life consists of immanent changes, behavior changes; also philosophical standpoints, even than philosophical axioms that seem external.

No research takes place in a philosophical vacuum [Hill, 1981]. Research is guided, consciously or not, by a set of philosophical beliefs; they influence or motivate the choice of topics and methods, and penetrate into the logistic treatment of the subject matter.

Is there a certain philosophy that best suits geography, regional geography in particular? Scholars borrow a philosophy which is in vogue during their time [Hill, supra cit.]. Sometimes, such philosophies have proved to be Trojan horses. The drift of regional sciences away from geography is one of the effects of this.

Geography explain the philosophical perception of its substantive subject-matter – *place, space, distance* – from which stem geographical paradigms. The very essence of geographical philosophy is [McDonald, 1966]. The generalisation of co-existent and co-varial phenomena as composites with areal expression. Geographical thought changed over time from Ratzel's anthropology, to Hettner's chorology, through Vidal's Human Geography, Sauer's landscape analysis, Hartshorne's regional geography, Harvey and Haggett's spatial analysis, to the Humanistic Geography of the late seventies, challenged by radical and marxist geography [van Paasen, 1984; cf. chapt. 1].

Along the course of its historical development, Geography devoted little attention to the epistemological study of itself [Gregory, 1978]. But in the seventies and eighties of our century, the epistemological-methodological-philosophical study of geography by geographers is one of the most active frontiers of research. The reasons for this may be an introspective insight into itself and a distant echo of the scientific occupation of most of the social and humanistic sciences in the last two decades. As Tatham [1951] stated, perhaps the most interesting aspect of the history of geographical thought is the sensitive way in which geographical ideas have reflected contemporary trends in philosophical thinking.

One opinion says that 'philosophing in an empirical science is a sign of trouble' [Couclelis, 1982]. This can not be true; a reflection on the way you are taking is quite normal in a world which changes at so hectic a pace. It seems to me that the current intense occupation with the basic philosophical and existential questions in geography is motivated by the hope of discovering a unifying perspective in a discipline beset by centrifugal forces which fragment the research, and different methodologies which deepen the gaps between the fellow scholars.

But the philosophy of geography is not only the perennial question, 'What is geography?'. Philosophy also means how a problem should be defined and what are the ways of solving it. Philosophy dictates the methodology, which is the logic of explanation [Harvey and Holly, 1981.]. Logic ensures that arguments are rigorous, that inferences are reasonable and internally coherent.

Habermas [cited by Couclelis, 1982] defined three approaches of inquiry, differing in their degree of penetrating into the nature of objects of study. The first, the primary level, is the *technical stage*: it is the step of fact-finding, empirical analysis, and is characterised by formaliztion of *statements* about covariation of events. The second level is *hermeneutic*: interpretation and explanation of the meaning of a text, of factual material, of the body of observations. The third level, termed *emancipatory*, is a critique and an examination of the inquiry itself; it seeks more general considerations and lessons, laws, that can be drawn from it.

Some scholars allow regional geography a 'raison d'être' as providing 'food for thought', seeing it as attaining only the first, technical stage, in order to provide subject matter to the two higher stages. Other, somewhat similar approaches to regional geography award it the status of a laboratory, where topical, 'systematic' laws can be tested. These approaches to regional geography see it only as a *ancillary* branch, which is, of course, indispensable but no more than other indispensable tools, such as paper or pencil. I do not accept this approach: the three levels of inquiry are *one unit*, each one a part of the whole inquiry. Geographical philosophing that is no anchored in reality is no less dangerous than dealing with facts without explaining them.

This erronously perceived, apparent dichotomy between the technical stage and the hermeneutic-emancipatory stages can be illustrated by a parable concerning two different approaches to knowledge. In ancient Hebrew literature, scholars are presented as metaphor, either as a 'living source' or as an 'impervious cistern'. The first one, obviously, has an original, innovating mind, producing new ideas, new points of view; the second one is distinguished by his knowledge, and by his ability to draw upon its depth. Of course, there do exist scholars who are more innovative and others who are 'treasuries' of knowledge. The ideal type is the scholar who both produces new ideas and possesses deep reservoirs of knowledge. Even a source – to return to the parable – is only a resurrection of water stored in deep hollows, caves and aquicludes. A true scholar is a source, whose water is preserved in a cistern that 'does not lose even a drop'. Good geographical research should comprise both the facts and their meaning; it should contain all three stages of inquiry, the technical, hermeneutic and emancipatory.

The prevalent philosophical trends of contemporary geographical thought are the positivist and the phenomenological schools [Harvey, 1979; Pickles, 1985]. The approaches of geographers towards one of these schools inhere more to their general education than to a conscious trend appropriated to the needs of the study.

Positivism is concerned only with tangible objects which can be quantitatively measured [Isnard, 1980]. It does not see ideological, religious, political or other mental element as objects of scientific study.

Phenomenology, although formalised at the beginning of this century [Lauer, 1965], penetrated into the geographical literature only at the beginning of the 1970s with a possible exception of Sauer [cf. p. 51]. Perhaps Samuels' doctoral thesis on existentialism in geography [1969], was one of the first works opposing the then-ruling positivist approach.

The phenomenological basis of geography had its turns in geographical literature several times; the most important being Vidal de la Blache's 'personality of the region', a concept rejected and attacked by positivists. Other phases were the 'immediate experiences of life' [Relph, 1977] and Buttimer's [1976] explanation of the French 'géographie humaine'.In the last decade, explicit reflections upon phenomenology have found their way into the literature, but according to Pickles [1985], these have been superficial forays. 'Phenomenology in geography has been much preaching and little practice' [Johnston, 1983].

Descriptive phenomenology is necessary to preclude abstract constructions and formulations. The goal of the phenomenological ontology is 'to return to the original data of man's experience'. Some important elements of phenomenology are appropriate to geography, and particularly to regional geography, for the basis of geographical thinking is geographical phenomena which must be explained. Husserl's watchword, 'Zu den Sachen selbst' [To the things themselves] relates to an approach to concretly experienced phenomena, a return to the immediate, original data of our consciousness; as we explained elsewhere [p. 61], a region is such an experienced set of phenomena.

Phenomenology suggests that

it is possible to obtain insights into the essential structures and the essential relationships of these phenomena on the basis of a careful study of concrete examples supplied by experience' and by 'a systematic variations of these examples' [Spielberg, 1974].

This approach seems to fit the approach to the notion of region: an example of a certain phenomenon which can be compared with other phenomena by studying the variations between them. We see here the justification of the study of geographical individual – regions – which are variables of a certain essence. Another element by which phenomenology is suitable as a philosophical basis to geography is that the positive data of experience are not restricted to the range of sensory experience, but admit on equal terms such non-sensory ['categorical'] data as relations and values ['We are the true positivists', Husserl, 1960].

It is important to understand that phenomenology is a critique of positivism, but not anti-science.

From a methodological point of view the credit given to subjectivity in research is a significant contribution:

... a subjective approach to objectivity is not illusory but rather the only approach which can prevent all objectivity from being self-contradictory [Pickles, 1985].

Phenomenology has now come to be seen as the philosophical basis in a number of the social sciences. It is a coherent and complete philosophical position towards the problemactics of the world. It is appropriate to the approach to the systemic region: phenomenology grasps the *essence* of the phenomena, when the essence is conceived of as that identical something that continuously maintains itself during the process of variation, the 'invariant', the universal and unchangeable structure. This is congruent with the approach to systems' processes as being influenced by different inputs, regulators, feedbacks and so on, while their essence remains the same.

Phenomenology contributes to education, by enhancing individual self-awareness, strengthening the sense of responsibility for the environment in which we live. Concepts of space, landscape, region have meaning to us, because we can refer them to our direct experience. With society governed by science and technology, phenomenological thinking serves as a humanistic balance in a technological world.

DIALECTICAL REASONING AS THE BASIS FOR THE CONCEPT OF SYSTEMIC REGION

The history of science or of philosophy reveals a controversy between claims for order and claims of scepticism [Szymanski and Agnew, 1981]. The world is ordered, but one must be sceptical about data and explanations, the purpose of which is to demonstrate this order. In a manner of speaking, order is thesis, scepticism is antithesis. Dialectics is reasoning which explains that every truth, understanding and explanation are only temporary and that they contain the seeds of their own transformation.

Dialectical thinking goes contrary to dichotomous thinking [Marchand, 1978]. As stated earlier [p. 7] the basic dichotomies in geography are only *apparent* dichotomies, in fact they can be bridged by dialectical reasoning [space/place, nature/ man, topical system/systemic region, etc.]. Dialectical thinking – *which is not at all synonymous with marxist thinking!* – is the optimal type of reasoning for dealing with the subject-matter of systemic regions.

Dialectical reasoning is a reaction to Cartesian reasoning. In recent centuries, science in the 'Western' world has been dominated by Cartesian reductionism and a mechanistic mode of thinking and explaining, which presuposes a fundamental distinction between parts and wholes, between causes and effects [Levins and Lewontin, 1984]. From this perpective, one of the goals of science was to discover the smallest internally homogeneous units of which the world is composed; to achieve this goal, more and more reduction was required, and the scientist was obliged to ever increasing microanalysis. But reductionism as *a universal road to truth* failed in many disciplines – as in the life sciences and the social sciences – because often an analysis which attempts to explain a phenomenon in terms of its microstructures loses the appropriate sense of cause and effect [by detaching it from its environment]. An active and productive alternative to reductionism is dialectical reasoning and explanation.

My purpose here – the examination of the notion of systemic region – entails a concern with the concept of the whole and its parts. In dialectic thought, wholes are [Levins and Lewontin, 1984]

a relation of internally heterogeneous parts that *acquire their properties by being a part of a particular whole*.

A holistic view is needed. Moreover, causes and effects are interchangeable and cannot be independently identified; sometimes the parts are the cause of features of the whole [Thompson, supra cit.], whereas in other areas, the whole is the cause of properties of its parts. This perspective was demonstrated through the example of holon [p. 24], and is an approach which describes a *more integrated* world than that described by Cartesian reductionism. As to methodology, the need to analyse the whole by fragmenting it into parts leads to tensions which must be resolved, according to dialetics, by synthesis.

Dialectics says that negation is also a positive fact, as it brings change, and change is characteristic of all systems. Time brings change, as everything is in perpertual transition into something else: although it may still be itself and so identified, at the same time it is becoming something else. A person, identified for a lifetime as a particular individual, changes from young to old; what remains constant is the phenomenological 'essence' [Marchand, 1978]. Through dialectics one can avoid the apparent dichotomies in geography, within which today exist extremes from mathematics to mysticism, by way of oposition, leading to synthesis [Marchand, 1979]. Dialectical synthesis can bridge two opposite, dichotomous concepts.

The differences between positivist and dialectical thought have been sum-marized by Marchand [1979] in a table, from which the following comparison has been drawn:

DIALECTIC THOUGHT	POSITIVIST THOUGHT
Qualitative analysis + synthesis	Quantitative analysis, no synthesis
Explicit structure beyond the appear-ances; subject/object	Accumulating of facts considered as external, given objects
Contradictions do not cancel each other	Formal logic
Quantitative change induces qualita-tive change; variables change, but remain themselves	Variables are fixed

METHODS IN REGIONAL GEOGRAPHY

Good geography begins with looking in the field.

J.F.Hart, 1982

Methods in regional geography include both those used in other branches of geography although perhaps [as with air photos and maps] in a more intensive manner and methods indigenous to regional geography [Nir, 1974]. This chapter will not be an survey of geo-statistical methods overall, but only mention those relevant to regional geography. It is assumed that the reader is familiar with statistical methods such as sampling, frequency distribution, probability, correlation and regression.

AIR PHOTOS

Vertical and other air photos have been among the most important tools of modern geography, since the second decade of the 20th century. Developments in techniques for reproducing and interpreting them led to a specialised field of 'remote sensing', promoted by the use of sattelites. The importance of air photos [of various types and technology] lies not only in their providing an objective picture of a certain area at a given time, but also in their ability to serve as a basis for a dynamic analysis of different phenomena, patterns and features through time.

The use of air photos has been one of the most important devices in establishing boundaries of different land-uses. To be able to use it, we must to make a preliminary field study identifying the expression of different land uses on the air photo; equiped with this key, it is possible to produce a map of the different land uses, based on the air photo. Comparison of air photos of a certain region, in discrect time intervals, allows us to study regional dynamics in land use – urban development, transport network, deforestation, changes in river bed, etc. The study of the meandering of the Jordan River [Schattner, 1962] could not have be achieved without being able to compare the pattern of the river in the decades of the 1920s, 40s and 50s by different sets of air photos.

Air photos can also be useful in studying social or human elements of a region. A census of a nomadic people can be seriously biased by the movements and dynamics of this population, with which routine census methods are of little use. Muhsam [1955] conducted a census of the Beduin nomad population in the Negev by using air photos to count tents. A preliminary field study of a random sample revealed that the average number of inhabitants of a tent is 5.6; by multiplying the

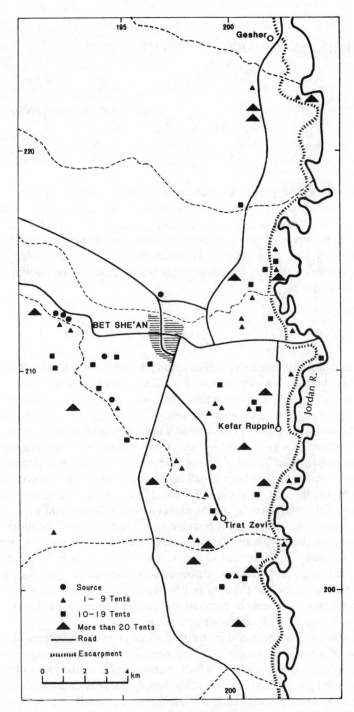

Fig. 13. Air photos as a method for enumerating the Beduin population in the Valley of Bet Shean, 1944 [Nir, 1974].

number of the tents in his photos by this value, Muhsam arrived at his estimation of the Bedouin population in the area under study. In 1944 we used [Nir, 1974] the same method for the Beduin population in the Beth Shean Valley [Fig. 13]. At the time of our study [1956/58] it was no longer possible to make a field study of the Beduin population in the region, and we used the Muhsam factor of 5.6 multiplied by the 536 tents appearing on the photos, and obtained a population of 3.016 people.

These remarks are only general considerations on this issue. Obviously, by using color air photos, satellites photos and transmissions, by computerizing of them, and by use of different artificial colors for different variables, and so on, a great deal can be accomplished toward obtaining a better understanding of the spatial distribution of the phenomena which interest regional geographers.

ANALYSIS OF MAPS

Mapping is the oldest and the most characteristic method in geography; *it is not only a method, but in fact a model*, abstracting, with growing precision, a certain reality. It must be admitted, however, that maps are loosing ground in favour of statistical treatment of geographical phenomena. Nevertheless, we still consider maps as both the basic method and primary expression of geographical thought; both the first step of the geographical process and an output of it, for in a map we express graphically and spatially the results of our research [Platt, 1935]. In this section, I wish consider the map as the primary source of information. Of course, geographical information should be supplied by statistical and literal sources, but the importance of the map is in its *synoptic presentation*, revealing different phenomena in their spatial interrelations. A regional research can use the map as the first step towards becoming acquainted with the region, even before conducting an on site field study.

The analysis of maps as a type of preliminary pilot-study of a region, the map being the only source, has been practiced in French universities [Tricart and Rochefort, 1953] since the beginning of this century. The maps analysed are at scales between 1 : 100.000 and 1 : 20.000, the optimum being obtained at the scale of 1:50.000; the area then is small enough to allow concentration on the elements and the relationship between them [topography, settlements, roads, land use, etc.]. In most cases, a geological map is also used, to facilitate the analysis of geomorphological elements of the region.

The aim of map analysis is not to quote the different and numerous elements of the region, not to repeat them by verbal description, but to reveal the relationships between the different elements drawn from the map itself without the aid of any additional information. Of course, the analysis depends on the quality of the map and the information included in it. Good maps include elements of land use, functions such as industry, institutes of education, types of religion, data on population, and can even be conducive to a preliminary economic or social analysis. This analysis should be considered a starting point for a study, a way

of determining the apparent affiliations of phenomena in the area and the relationship between the relief, hydrography and other environmental elements and the
implications of human activity on them. Analysis of a map gives the basic information on the inventory of the region; although not a sufficient one, but it is
nevertheless a necessary condition to conducting a study. The analysis of a map
should be guided by some basic principles:

1. *Orientation*: [the map in the frame of a smaller scale] showing the connections
of it with neighbouring regions, its affinities with a district, township, etc.

2. *An essay to divide the map into subregions*, which can later be treated separately.
The criteria for such a division should be those most relevant to the character of
the area on the map, be it of topographical or socio-economic nature. In a
mountainous country, perhaps topography will be the best criterion of division;
in a highly populated region, perhaps the pattern of settlement; in an urban area,
division into functional subregions; in an agricultural realm, different land-uses
such as fields, pastures, woods, etc. There is no uniform criterion, no universal
'scheme' [p. 89], as in each region – and a map is a graphic abstraction of a region
– different elements are arranged in different relationships.

3. *An attempt to see the man-built system within the natural challenges*. This attempt
will confront environmental elements with element of human occupation and
activity.This confrontation will perhaps show not only a static situation, but
developmental stages of relationships: roads becoming more and more inde-
pendant of topography; a town developing from a nucleus conditioned by a certain
element – a bridge on a river, perhaps – in a different pattern according to new
trends, etc. Perhaps some unexpected problems will be evoked by this confrontation. A full answer can not be obtained from the map alone, but the importance
of the primary approach to and the analysis of a map lies not in the answers it
provides, but in the questions it poses.

4. *Analysis of the anthropo-geographical elements*. We have to keep in mind that
regional geography is a consideration of organisation of societies in a given
environment. This organisation can be traced on the map by the type and size of
settlements, by the infrastructure of them and by economic activities, which can
be traced on maps of approppriat scale [factories, types of residence]. The
existence or absence, predominance or rarity of certain elements can give us a first
impression of the character of the particular region.

Gathering together all the information and questions, we can arrive at a preliminary impression, which will be not more, but also not less, than a 'working
hypothesis', to be attacked by other tools and methods.

Air photos and maps are methods more typical of regional studies, and perhaps
not so much used in other fields of geography. Other field methods of regional
study are common to most if not all branches of geography and social studies: the
use of literary sources, archives, statistics, questionnaires and interviews. I shall

not go into these methods, as they are familiar to the student of geography from his basic training.

DELIMITATIONS OF BOUNDARIES

Delimitation of a region and establishment of its boundaries have often been presented as the basic task of the regional geographer – which, of course, they are not. Incidentally, if a region is a real entity, it is plausible that it will possess definitive boundaries; for some critics of the regional concept [Grigg, 1967] the absence of clear delimited boundaries of a region was perceived as negating its reality. However, the treatment of the definition of boundaries has been quite exaggerated, and more intensive than the treatment of the proper nature of the region, especially when defining of boundaries could be used to camouflage political aims [p. 19]. Of course, a definition of boundaries of a region is essential, and an effort should be made to find relevant criteria for defining them according to the nature and essence of the region. The problem of boundary definition for a functional region or in a one characterized by a single common variable, is different from that for a multiple variables region, or the systemic region. Whereas in the first case a linear boundary can be established, the later two types of region would be delimited by a more areal boundary, that is, by a transitional area. This method of boundary delimitation in no way reduces its reliability.

Boundaries of regions having one common variable

When the quality or variable characterizing an area is of a homogeneous distribution, the boundaries of it are self-explanatory; no special method is needed for their elaboration. So, if we are delimiting 'plain' from 'hill', and we define a 'plain' as anything lower than 200 meters in altitude, and 'hill' as anything above 200 meters, then the isoline of 200 meters is the boundary between these two variables. In such a case, there is no place for ambiguity. The same is true if an area is delimited by pedological criteria [terra rossa/rendzina], by land use [field crops/ orchards], population density [less than 100 inhabitants per sq km/more than 100 inhabitants per sq km], etc. The limits of a variable, or an amount of a certain quality expressed in numeric value, are the boundaries of the area it occupies. As explained elsewhere [p. 63], this type of area is defined as a zone and not as a region.

But a definition of the boundaries of a single-variable area becomes problematic if this variable is not continuous, or if its distribution is conditioned by seasonal or annual changes. Consider the first case – discontinuity in space – in which, for example we wish to delimit distribution of two different land uses in an area [irrigated crops/dry farming; residential area/commercial area]. Each plot of the particular land use is defined within its own limits, but to characterize an entire area as being irrigated and to delimit it from the dry one, or to draw a limit between the residential and commercial quarters of a city, requires an operation. Consider a value of a variable which is continuous in space but not in time – isohyet of a certain value of rainfall, or a value of income per capita, etc. – if we wish to present this value as a multi-annual limit of its areal distribution. It comes out again that there an operation is needed to define from the individual limits of plots or of individual values a representative boundary. This can be done either by establishing an arithmetical average or by using the median of the distribution [Nir, 1974].

a. The use of arithmetical average

Generalization is a method used in every scientific procedure. It should not be considered a biased method; the reliability and significance of a generalisation depend upon the definition of the desirable exactness. If we decide that an area 75% or more of which is irrigated will be defined as 'irrigated', and that an area 75% or more of which is unirrigated will be defined as 'dry farming', and that an area more than 25% irrigated and less than 75% will be labelled 'mixed cultivation', then we have a generalization which is exact in its definition, but which, on the other hand, leaves a large intermediate area. This exact delimitation of the distribution of the variables does not greatly distort their real distribution; we do not, however have a clear delimitation between the irrigated and unirrigated areas, as we created a large, perhaps over large intermediate area. If we wish to obtain a delimitation which is more clearer and more intersecting, but more generalized, we can increase the limit between the two variables. We can even decide that the demarcation between irrigated and dry farming is the break between 49% and 50% of irrigated plots. The result is an exact line, but its information is less reliable, as such as area, although defined as irrigated, can contain up to 49% of unirrigated parcels. But as long as we define, a priori, our criteria, we can trace both a line and an intermediate area. It will depend on the scale of our investigation: which of the two methods to use [exact line or intermediate area]. The larger the intermediate area, the less significant it is for the purpose of delimitation. On the other hand, the width of such an intermediate area is important in and of itself, as an indicator of a rather diffuse distribution of the variable. Variables to be delimited by an areal rather than a lineal method are mostly from the socio-cultural and economic realms, such as distribution of languages, ethnic groups, land-uses, and so on.

b. Delimination as the median distance between extreme occurrences

This procedure can be applied to variables which are related to seasonal, annual or periodical changes, when the variable depends on a constant factor, or if it changes over time. Variables may derive from the either social or environmental realms. An example is the demarcation of aridity in the central part of Israel [Ganor, 1963]. Figure 14 shows the yearly limits of aridity from the years 1940/41 to 1959/60. In the year 1959/60, to the north of the limit of aridity only Mediterranean type of climate occurred; In 1944/45, only arid climate occured to the south of the limit of aridity; in between, on a strip 60 to 80 kilometers wide, both mediterranean and arid conditions have been registered. Therefore we have

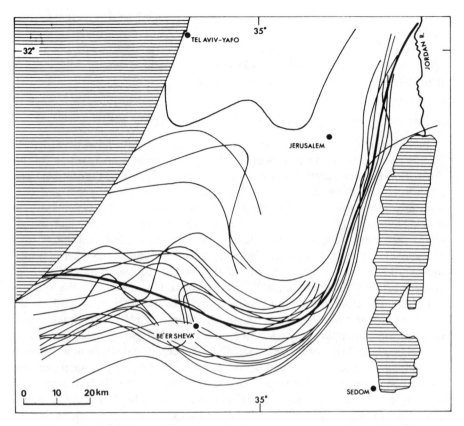

Fig. 14. Delimitation of the median distance between extreme occurrences: limit of aridity in the central part of Israel [Nir, 1974].

here two major climatic realms, and between them, a semi-arid climate. If we wish to delimite arid and humid climates, we must find a way to draw it within the intermediate realm. An arithmetic average of the distance between the extreme lines of aridity does not express the reality, as the rare extreme shifts to the north and to the south should not to be disregarded here. It is rather proposed to use the median and not the average, i.e., the limit will be the line which has the central position between the extremes. The heavy line on Fig. 14, which is the median of all the lines, is considered the boundary between arid and humid climates. This line says that to the south of it, in 50% of the years of measurement the climate had arid characteristics, whereas to the north of it, in 50% of the years it had mediterranean characteristics. Perhaps the best procedure is to present both the

limit of aridity and the intermediate semi-arid area, as it is useful to present average
and median values with their deviations.

Which of the boundaries, lineal or areal, we will use, depends on our scale –
whether we are dealing with a world map of climatic zones, or with local or
regional problems and with drought in Israel.

Delimitation of the multi-variables region and the systemic region

A multi-variables region and, especially, a systemic region have been defined
[p. 66] as an interaction of elements existing in a certain area, which by this
interaction become components of an open system. The more they interact and
interweave, the more the area becomes different from other areas, and consequently, a systemic region. The more the processes acting in the systemic-region
become increasingly interconnected, the more the qualities becoming increasingly
similar, the more congruent the delimitations of these qualities will be. A total
congruency can exist, however, only if the variables are linked together in a total
manner. This is not to be expected, for that would require an impossible congruency between the environmental, social, economic and iconographic qualites
of a place. Therefore, the definition of boundaries of such regions depends on the
degree of *partial congruence* or correlation between the particular variables and
qualities. In a multiple-variables region or a systemic-region, we have to accept
a priori that within the core of the region, the connections and interactions between
the variables will be stronger and within another, the periphery of the region, the
connection and activities between the variables will be weaker. In fact, a structure
of 'core and periphery' is quite common in most realms, be they political, social,
economic or climatic, and this is especially true in systems of civilisations. It must
be admitted that there exist boundaries of a certain type, which are clearly and
exactly defined,and which also have an exact delimitation of their functions: these
are the political and administrative boundaries, which are decisive to the nature
of the regime, legislation, economy, and civic life within them as distinctive and
different from those existing outside them.

Generally speaking, if a region is characterized by a certain interaction between
its components, its boundary should be traced where this interaction ceases. In
most of the cases, this will be administrative or political boundary, because social
activity is strongly influenced by administrative and political realities: we are
influenced by laws, national economy, state or municipal decisions which have
influence us on within these political or administrative boundaries.

Let me illustrate the case of political boundary, or how a boundary of a single
variable can become a boundary of a multiple- variables region. The example is
the boundary between California, and Mexico. At the time it was established, the
political boundary between the two nations, an arbitrary, more or less straight line,
had only a political meaning. The landscape and the land use on both sides of it
did not differ greatly, if at all. It was a political, not a geographical boundary.
Needless to say, this one-variable boundary became a true geographical multi-
variables boundary, where the population, landuse, landscape, nearly every aspect

of social activity on one side of it was different from on the other one. As a consequence of the different political and economic situations on either side of the boundary, it became a boundary between systemic regions, delimiting irrigated realms, cities of millions, and a highly developed road system on one side, from a sparsely populated area on the other side. This example illustrates that even in a systemic region, *one* element can be sufficiently decisive to trace the boundary, the other components of the region being implications of it.

The less congruency there is between the different variables of a region, the more difficult it becomes to establish the boundaries of the region. No axiom states that boundaries of *all* the variables of a system-region should be congruent. In the parts of the region delimited by congruent boundaries, the interrelationships will be stronger than in the parts where they are not, but the latter still belong to the region, because of the basic links within in them even if they interact at a lower intensity. Figure 15 shows different limits constituting the boundaries of the region of Bet Shean: international boundary, administrative boundary, limits of so-called 'natural boundary' of the valley of Beth Shean. The boundaries are not congruent; but the congruency lies in the regional interacting, as each village participates in the municipal-regional activities and in the regional economic activities through the region-wide cooperation in water use, in processing of agricultural products, etc. The municipal boundaries of the Regional Council are in fact the boundaries of a systemic-region, as all the economic, cultural and municipal services are carried out under its common denominator. The differences in relief – valley, plateaux, hills – although important, are not decisive in understanding the social and human originality of this region.

The kibbutz Sde Nahum [Fig. 15] is situated in the north western extremity of the region of Bet Shean; some five kilometers to the north-west of it lies the kibbutz Bet Hashita. The two kibbutzim are situated in an identical environment: on the line between the inclined Plateau of Zvaim and the Valley of Harod, on the same basaltic red soil, having the same climatic conditions, on the same highway, affiliated to the same kibbutz organisation. But administratively Sde Nahum belongs to the municipality of the Bet Shean Regional Council, whereas Bet Hashita belongs to the municipality of the Harod Regional Council. This administrative boundary, not to be observable in the landscape, is decisive: Sde Nahum participates in the cultural, economic and municipal activities of the Region of Bet Shean, whereas Bet Hashita regional activities are focused around another pole. Thus, although from the environmental point of view, Sde Nahum and Beth Hashita are closed to one another than to any other villages in their vicinity, it is political boundaries which are ultimately determinants the interactions between the villages in the Bet Shean region, and not the environmtal similarities. The postulate that in a systemic-region the boundaries of all the variables should be congruent, and if this is not the case, the region is not an entity, is inappropriate. A region begins and ends with mutual interaction, mutual problems, mutual answers, mutual challenges. Where these cease to exist, the region ends.

An illustration of a region delimited by the different functions of a pole – a city – superimposed on its surroundings is the regional influence of Natanya, a town

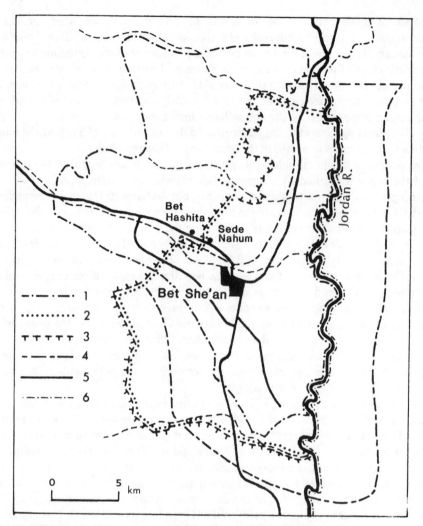

Fig. 15. Different criteria for a delimitation of the region of Bet Shean: 1. The limit of the Beisan Sub-district, 1920–1948; 2. Limits of the Regional Council of the Bet Shean Valley, 1948–1982; 3. Dito, since 1982; 4. Limits of the physiographic unit Bet Shean Valley; 5. Roads; 6. Israel–Jordan boundary [Nir, 1974].

in the Coastal Plain of Israel. The variables included in the study are the boundaries of the municipality, of the police district, of bus services, of gas distribution and of medical services [Fig. 16]. The sum of these services creates an area having a core where all these variables are present, and a periphery where not all of the functions are represented. This procedure can be a basis for investigating the pattern of the boundaries of the region of Natanya.

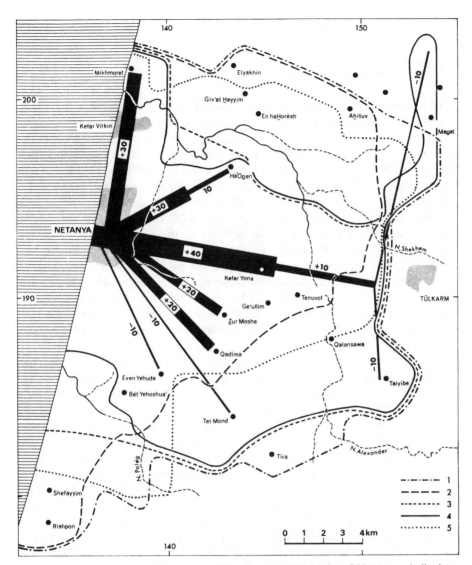

Fig. 16. Regional influences of the town of Natanya, Israel: 1. Limits of Natanya sub-district;
2. Limits of Natanya's police district; 3. Limits of the gas-company; 4. Limits of the district's health's
services; 5. Limits of the local bus-station [Nir, 1974].

DEALING WITH VARIABLES

The importance of personal choice

The way in which variables are chosen, selected or defined, is one of the touch-stones of research, and certainly of regional geographic research where the varia-bles are, theoretically, the totality of the region. Irrevelant variables obscure the results of a research [p. 83]. On the other hand, omission of a variable having specific significance can bias the entire study. There is perhaps no other step in regional geographic research so crucial as the choice of research variables and it requires the ability to discern the essential from the incidental.

Choosing the variables for a study is a test of the scholar's ability and in this part of the research, the subjective qualities of the scholar are decisive: his education, skill, aims. Even with the best intentions of scientific objectivity, i.e., without prejudice or extra-scholarly tendencies, the socio-psychological background of the scholar cannot be eliminated from such issues as the choice of a project of study, or the choice of variables relevant to the study, because relevancy is weighted by the scholar's decisions, and these are part of his per-sonality. We must accept that the choice of variables is a personal affair [Cole and King, 1968].

As to the relevancy of a variable, we proposed [p. 83] that by adopting the systems' approach, we must ask in deciding whether or not to include in a study, if the particular variable seems to be connected with others, and if it contributes to the systemic region. If the answer is positive, on a prima facie level, it must be included in the research. As the decision is of elemental importance, we should be careful about withdrawing of a variable. The best method is to conduct a pilot study [p. 125] to test the main problems of a region, and then relate the variables to its results. We should not eliminate any variable without testing it by such a preliminary test. Sometimes a single variable can modify the entire situation.

The importance of a single variable is illustrated by a study on cattle raising in South Africa. At the beginning of the 20th century, South Africa seemed to have good conditions for cattle breeding [Talbot, 1961], provided there was a sufficient water supply in the north-western semi-arid part of the country. Notwithstanding the good supply of water and relatively rich pastures, however, the mortality of the herds was high, for no apparent reasons. A pedological research – an element that seemed unnecessary, as the conditions for breeding were water and pasture – revealed that the soil, and therefore the pasture, was poor in nitrates, being a product of the erosion of crystalline rocks. The herds suffered from the absence of nitrates, which became a hazard to their survival. Supplying nitrates saved the herds. Thus a single, neglected element can be of crucial importance in a sub-system [soil], which influences another sub-system [natural vegetation] which is directly linked to the principal system [cattle raising in South Africa].

There are single variables not indigenous in a region but part of the external input [p. 91], such as political decisions of the government, technological innovations from abroad, etc, which are 'hidden' or obvious to the observer. An example of such a unique factor is drawn from the development of the Plain of Sharon in the Israeli Coastal Plain, between 1850 and 1950 [Karmon, 1959]. The change in the landscape of Sharon during these hundred years is comparable to the changes which occurred in the lowlands of the entire Mediterranean Basin, from a nearly abandoned region to the most densely populated and highly developed agricultural zone in the country. Karmon revealed that the primary factor enabling development was the ability to penetrate by drilling to deep lying water tables and to pump the water from the depths to irrigate the red-sand hills. Since Creation the red-sand hills had been only pastures, woodland or grassland. Now, thanks to irrigation, they became thriving citrus orchards. The water, the climate, the soil, all were inherent; but the technological achievement – deep drilling and engine powered pumps – was the key to the revolution in the relationships between man and environment in Sharon, and in many other agricultural regions in the world. Of course, this revolution would have been unthinkable without the economic and cultural background of the population. Explaining the development of Sharon without giving credit where due obscures the understanding of the processes that gave to the region its current landscape.

Certain variables no longer active today, were active in the past, and they should be brought out in a regional study, as under certain conditions they can return. At the beginning of the twentieth century, Jerusalem, at an altitude of between 700 and 800 meters, on the fringe of the desert, remote from rivers, lakes and swamps, was affected by endemic malaria, induced by neglected water cisterns in the city. Of course, no scholar will look into malaria as an environmental element of present-day Jerusalem; but a scholar dealing with Jerusalem at the beginning of the century, can not dismiss, as it was a component of the regional system. A scholar studying a region must *discover* the variables that are not visible on the surface.

Isolation of the 'central problem'

The choice of relevant variables to be included in a study should be focused on the problem or problems which characterize the region [p. 91]. It is not possible to deal with all the elements in a systemic region. Our discernment of the essential from the incidental can be compared to the reasoning and procedure of a medical general practioner examining a patient: he must know the general condition of the body [and of the mind], but reaching a diagnosis, he must concentrate on the illness which has been found to be acute. This is especially the case in studying a region with no prexist hypothesis and no aim but to discover its problems.

On the other hand, the decision to study a particular segment of the globe can be stimulated by a particular interest in that region, a pragmatic reason of organisation or planning. In each case, we need to deal with variables of vital importance for understanding the region. To discover, the 'central problem' we must begin with a *pilot project*. The main research will be devoted to checking and studying of the problems revealed by the pilot project, while other variables in the region will be considered, as compared with the central problem, banal and irrelevant to the study.

As suggested above, the pilot study can be made, in either of two ways: by the 'tabula rasa' approach, or by the approach of a preliminary hypothesis.

The first approach entails gathering data and information on variables in the region, with no initial preferences for any from among them. By weighting and discussing the variables and their interactions, we may discover, perhaps even at an early stage of investigation, the specific importance of a certain element or group of elements and the differences between the region under study and adjacent regions. The trend should be toward an investigation of the variables responsible for that difference. If the differences continue to be significant and relevant, it should be considered the central problem of the region or, at least, one of the central problems.

If we approach a region with a working hypothesis, we must check it against the actual variables from all possible angles. If this procedure affirms the validity of the hypothesis, it will be considered the central focus of the investigation; if it is not affirmed or if it is refuted by the problem, it should be abandoned and a new hypothesis on the 'central problem' chosen, according to a general consideration of the variables in the region, as mentioned above.

In most cases, a study of a systemic region reveals that even in a region where the approach was not based on a preliminary working hypothesis, there exists a central axis of problems or issues, which are more important than other variables. It is natural that these themes will become the axis of the research.

The designation of a variable or group of variables as the central problem of a region, as determined by the pilot project, need not be identical with the final results. These can be reached only as issuing from principal research. The principal research may even annul the importance of the variables considered from the pilot project as the central problem of the region. The 'central problem' remains a working hypothesis until it is either approved or dismissed by the principal research. As an example, I shall return to the study of the region of Bet Shean.

A pilot project quite soon revealed that the variable that distinguishes the region of Bet Shean from other segments of that part of the Jordan Valley is the presence of rich sources. Springs at that level of discharge – 90 million cb.m./year – are virtually unique in Israel, aside from the sources of the Jordan and the Yarkon Rivers. Therefore, the preliminary research focused on the springs and their consequences, although the problems of soil, climate, distances from the center of the country and so on, were not neglected. The economic-ecological importance of the springs makes them the central issue of the region. In fact, the springs are likely to make that area a region because they are, in a semi-arid environment, the dominant element in the economy of the population.

But the basic research revealed that the occurence of the springs per se is not the 'main problem' of the region; it is rather their use or misuse. Not in all periods of history the region has been so prosperous as it is today. The Roman-Byzantine period was one that of thriving urban and rural settlement [Nir, 1968], but the late Middle Ages and the Modern Period until the very end of the 19th century were periods of waste and misery, trachoma, malaria and depopulation. It must be acknowledged that not the springs themselves but their use by the local population is the 'central problem' of the region. Indeed, the outcoming basic research did focus on this. Isolation of the central problem is only the end of one stage; the development of a theme through a basic research refines the approaches to the central problem and results in a satisfying answer to the questions asked. The importance of the pilot project is in *focusing attention* on the problems of a region.

Each 'central problem' of a region is relevant only within a certain timeframe. As with all study of human behaviour and action, it is conditioned by the historical situation. Our research of the region of Bet Shean was conducted in the late fifties and mid-sixties of this century. With our renewed interest in the region in the early eighties, it became apparent that the 'central problem' of the fifties and sixties was no longer relevant. Rational and economical use of the springs, based on modern and constantly improoving technology, which was the base for regional organisation and cooperation, became routine, a part of the establishment. But new problems had arisen in the twenty years since the previous investigation: the town of Bet Shean, which in the late fifties comprised some nine thousand inhabitants, became a town of fourteen thousand inhabitants in 1986, comparing with ten thousand inhabitants in the rural sector.

Moreover, the 1950s population of new immigrants in the town became in the early eighties an emancipated population. Because of political differences between the urban and rural sectors, as well as socio-economic differences between them, a certain friction exists between these two populations. Although, of course, any recklessness in managing the springs could have disastrous consequences, the 'central problem' now is the relationship between the urban and rural populations. The 'central problem' is not a static concept, but the *constantly renewed questioning on the nature of a region.*

DISTRIBUTION OF VARIABLES

Relationship between variables and area

The study of the distribution of variables in an area is one of the bases of regional geography, and it contributes to the study of the relationship between different phenomena. The first step in studying the distribution of variables in an area is their enumeration and description. This descriptive, documentary, primary and necessary step does not enter into the analysis of the distribution. The scholar contented with reaching this step cannot lay claim to doing scientific work. A map gives an inventory of an area; acquaintance with inventory is a necessary step, because processes are based on facts. But a graphic representation of the distribution of a variable leaves open the judgement of its quality. As this differs from person to person, it is not wise to leave an *evaluation* of this representation to visual impression.

One of the basic representations of a distribution in an area is that of the relationship between the variables in question and the size of the area – the *densities*. This rather well-known parameter of distribution measures the intensity of the relationship between a certain variable and area by dividing the numerical value of the variable by the area [in square miles or kilometers]. This parameter can be compared with values of the variable in other areas; different areas can be compared and treated.

The distribution of variables in an area can be measured by their relation to a certain point of departure, from which distances and densities can be established; these are the coordinates that make possible a quick transformation from geographical coordinates to a cartesian system of distribution. The scale used in this procedure can bias the visual impression, showing great densities [small distances between the variables] or small densities [a greater dispersion between the variables]. It happens that a graphic representation is biased according to the scale used and can mislead if used improperly. But if we represent the distribution of variables in a numeric procedure, this bias can be eliminated. Therefore, a numeric treatment of variables has to complet the cartographic one.

Random distribution and distribution in clusters

Distribution of a certain variable – farms, corn fields, trees – can be at random, meaning that the distribution is not conditioned by a secondary factor, or it can appear in clusters, with some reason behind this. Equal ecological and economic conditions can produce a regular distribution, whereas particular conditions can produce a distribution in clusters [Zobler, 1957].

The measurement of distribution in such cases begins with parcelation of the area into smaller rectangular areas, fixed on axes of X,Y, or longitudinal and latitudinal coordinates. In these arbitrary areas the individual items of the variable are enumerated, and the relations between the areas examined by the X square test. An example: we must examine the distribution of oak trees in a woodland of mixed forest, in an area of 9 hectars. The number of oak trees in the area is 45. We wish to determine whether the trees are dispersed at random, with no local factor influencing their distribution, or if there exists some local factor.

The area of 90,000 square meters is parcelled on an air photo, into a number of smaller squares. In this case, into 9 parcels of 10.000 square meters each. If the distribution of the oak trees is be equal, we can expect in each quadrangle five trees; this is the *expected value*, E. The number of oak trees actually existing in the quadrangle is the *observed value*, O. The X square test is

$$\text{X square} = \frac{[O - E] \text{ square}}{E}$$

where E is the expected and O the observed values.

Fig. 17 shows the actual distribution of the oak trees in the area under study; Table 1 shows the treatment. The result is that X *square* = 38/5 = 7.6. In the

Expected				Observed		
5	5	5		3	3	2
5	5	5		6	8	8
5	5	5		6	5	4

Parcel	O	E	O − E	$(O - E)^2$
1	3	5	− 2	4
2	3	5	− 2	4
3	2	5	− 3	9
4	6	5	+ 1	1
5	8	5	+ 3	9
6	8	5	+ 3	9
7	6	5	+ 1	1
8	5	5	0	0
9	4	5	− 1	1
N = 9	Σ = 45	Σ = 45	Σ = 0	Σ = 38

$$X^2 = \frac{38}{5} = 7.6$$

Fig. 17. Expected and observed values, X square test.

probability table of square X values, in 8 degrees of freedom, we find a probability of less than 0.50 that the difference between the observed and expected values can be at random. The distribution of the oak trees seems to be conditioned by an until now unknown factor. We shall, therefore, undertake an investigation into the factors causing this distribution.

Center and periphery of distribution

Parameters of densities of distribution of a certain variable relate to the whole area in which this distribution is measured. The larger the area in which the distribution is measured, the more the actual distribution in a concret place can be misrepresented by the average value of the distribution, as it is representative of the whole area. So, the average distribution of the density of population of a nation or state need not represent the density of a county or district. The average density of Israel [1.1.1986] is 196 p/sq.km, but the density for the districts is quite different from that: in the Northern District it is 157 p/sq.km., in the Southern district [including the Negev] only 36 p/sq.km., whereas the Central district has a density of 715 p/sq.km., the Jerusalem District 807 p/sq.km. and the Tel Aviv District 5972 p/sq.km.

For each average, the standard deviation is a value which should accompany the data. We can establish a distribution of a certain variable – say population – for a county, district, state, nation, even for a continent, but the actual density of a small area will be always different from the average density of the total one. In most distributions, we must distinguish between their core and periphery. One of the parameters allowing us to discern between the core and the periphery, is the parameter of center of gravity of a variable.

Center of distribution of discreet variables

This study of distribution is concerned with finding the center of distribution of discreet variables having the same value [Cole and King, 1968]. The area under study is divided into rectangular Cartesian quadrangles, on X and Y axes. The elements of distribution are collected along their X axis and weighted with the values of X; the resulting number is divided by the number N of the elements of distribution, and a value for the central place for the elements on the axis X is obtained. The same will be done with all the elements according to their distribution relative to the Y axis, and an average position for Y will be obtained. The two average values of X and Y will intersect in one point, which is the *center of distribution* of the variable under study. This measurement of distribution is possible when each discreet member of the variable has the same value, i.e., when each one is one unity.

Center of gravity of area

The center of distribution of a discreet variable, as defined above, is not necessarily the center of the area in which this variable is distributed. Indeed, one of the most interesting questions in regional geography are the relationships between the center of a distribution and the center of an area [region, state]. A region can have a regular distribution of a certain variable [population] or a irregular core/periphery distribution, and the study of the reasons for it can be quite fascinating.

The method for calculating the center of gravity of an area is identical to that for the center of distribution, but the axes X and Y are the geographical coordinates of the latitude and longitude [Bachi, 1957]. A central value of longitude which traverses the region [county, district, state] will be chosen, and the values of the latitudinal axes that traverse the region will be be counted, giving at each degree, or 30 minutes, or any choosen interval of latitude, its longitudinal value. The same will be made with the longitudinal values on a latitudinal axis. We will establish the Y values for the longitudinal axis, and X values for the largitudinal one; the resulting values of X,Y, intersecting in one point, will be the center of gravity of the area of that region, county or state. This procedure is possible, because each value included has only one quality – its distance from the X,Y axes, expressed in kilometers or miles.

Center of gravity of a population

Up until now have we dealt with distribution of identical variables [trees, kilometers]. In the next section, the problem posed is to find the central point of the distribution of a variable having different values. The difference between this procedure and the previous one is that we must weight the local value of the variable [coordinates] with its quantitative value [number of variables on each coordinate]. The most common theme is the distribution of the population of a region. It is impossible to mark by coordinates the exact position of each inhabitant of a region; but the identification of the site of his habitation is possible. The analysis of the center of gravity of population of a region is built, therefore, on weighting of a site [coordinates] by the number of the inhabitants of the site, at a certain time. The formula is [Bachi, 1957; Statis. Yearbook, Israel, 1986], for the latitudinal coordinate of the center of gravity X:

$$X = \frac{\epsilon \, p_i \cdot x_i}{p_i}$$

when p_i is the value of the variable p [population] in site i, and x_i is the coordinate x in site i. In a similar way, the longitudinal coordinate of the center of gravity Y is established:

$$Y = \frac{\epsilon \, p_i \cdot y_i}{p_i}$$

The definition of the center of gravity of the population in a region is the point, the coordinates of which are the averages of the coordinates of all the inhabitants of the country. This is the point where the total of the squared distances from it to the particular site of habitation of each inhabitant is minimal [Bachi, 1957.]

The center of gravity of population is an important tool which gives immediate information on how the population of a region is distributed. Australia [Fig. 18] has its center of gravity of area firmly in the middle of the Commonwealth, whereas its center of gravity of population, situated in the south-eastern corner of the Commonwealth, reveals the asymmetry of its distribution. India, to illustrate a quite different distribution, has its center of gravity of population in its center of gravity of area, which hints at a more weighted distribution throughout the country [Fig. 18].The two different situations are crucial in communications, commerce, and practically all aspects of the social activities of the two nations.

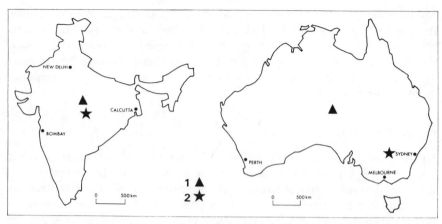

Fig. 18. Centers of gravity of the area [1] and the population [2], Australia and India [Neef, 1966].

The center of gravity of population can also be a tool in studing the dynamics of movements of populations, the trends of its development and changes in the population map of a region, if applied in different time periods. In the region of Bet Shean, which is our case-study throughout this book, the centers of gravity of populations in the years 1948–1982 were rather stable [Fig. 19], indicating by their positions a rather weighted development of the area in this period. The center of gravity of population was almost in the center of the area of the region; new settlements established during this period, to the south west and the north-east of it, counterbalanced each other. The importance of the urban population is revealed by the fact that the centers of gravity of both rural and urban populations are always situated in the urban area.

The trends in development of the population distribution in the US are very convincingly represented by the centers of gravity of population in the two hundred years since 1790 [Fig. 20]. There is a steady movement of the center of the gravity of population from north-east Maryland in 1790 to the west, reaching Cincinnati a hundred years later, and situated in the eighties of this century near De Soto in Missouri. The figure also shows, that the trend until 1940 was almost directly westward. Since 1940, it has become a westerly trend with a southern component. The advancement to the west was enhanced between 1810 and 1890, as shown by the decennial distances between the centers; the pace slowed between 1910 and 1940. Between 1940 and 1980, the advancement of the center of gravity of population became anew very rapid. It clearly reflects the migration boom to the south western states, especially to California, in the post-war years; the advance being 250 km to the south-west. The center of gravity of the area of the US is situated on the border between Kansas and Nebraska, some 500 km to the north-west of the center of gravity of population. The trend in these two hundred years was from the areal periphery in the extreme east to very near to the center of area of the nation, which means, that today the areal distribution of population of the US is nearly balanced.

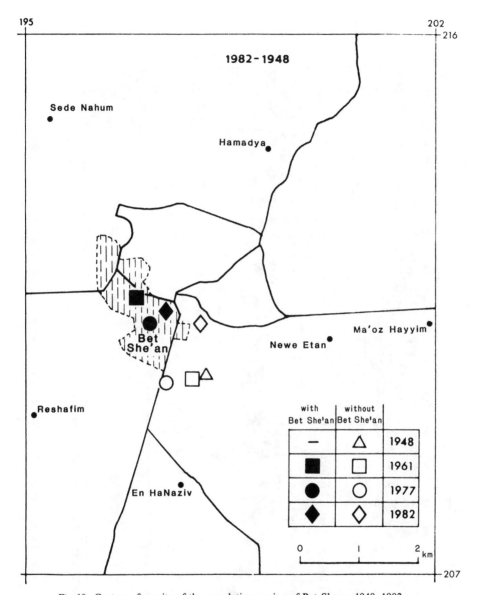

Fig. 19. Centers of gravity of the population, region of Bet Shean, 1948–1982.

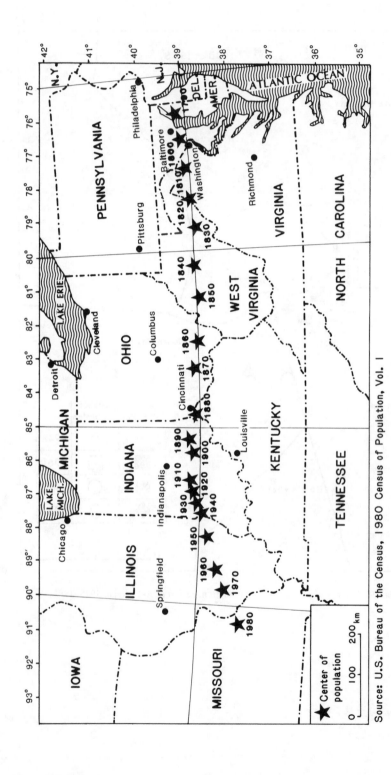

Source: U.S. Bureau of the Census, 1980 Census of Population, Vol. I

Fig. 20. The movement of the center of population of the US 1970–1980. U.S. Bureau of the Census, 1980 Census of Population.

TOPOLOGY

Topology makes it possible to deal with regional geographical phenomena by an approach different from Euclidian geometry. The aim is not to consider exact distances and directions, but the relationship between the phenomena. This enables the study of relationship without diminishing the credibility of the graphic representation. I shall illustrate the essence of geographical use of topology by the example of a topological representation of a part of the regional Metro in Paris, between the stations St. Remy and Luxembourg [Fig. 21]. On the left, is a

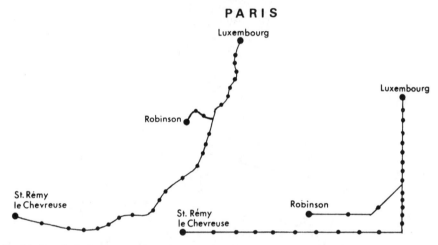

Fig. 21. Topological representation of a part of the Paris Regional Metro, between the stations St. Remy and Luxembourg [Nir, 1974].

representation of the line, actually a part of a topographical map, where the distances and directions on an appropriate scale according with those existing in reality. On the right is a topological representation of the same line, where neither the exact direction nor the exact distance between the stations is represented, only the sequences of the different stations and the ramification of the line. The topologic transformation is, first of all, pragmatic. It removes the line from the complexity of particular curves and sinousities and differences in distance between stations, and gives the passanger the essential and only relevant information concerning his passage: the sequence of stations and the ramification of the line. Additionally, the names of stations can be more clearly identified on a nearly straight line than by the sinuous topographical presentation. We can define the topographical transformation as a presentation of a phenomenon according to a certain aim, here, an unmistakable representation of the sequence of stations, by neglecting the [here] secondary qualities of distance and accurate directions of the line.

This representation can be a basis for comparisons of linear features, such as roads or rivers, when the exact shape of the rivers is straightened, allowing a

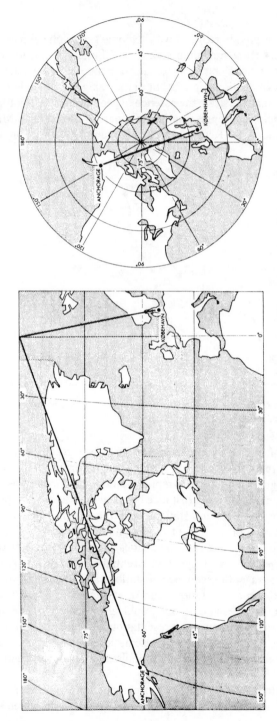

Fig. 22. The distance København–Anchorage as represented by different projections [Nir, 1974].

comparison of the number and lengths of its tributaries. Most transportation maps are topological representation of flow of commodities or people, where the interest in the main variable can be stressed and the secondary elements neglected.

Maps are, in a certain measure, an expression of the topological approach, but not all maps are suitable for topological transformation. Mercator's projection keeps the true direction from one point to another one, and is therefore suitable for navigation; on the other hand, it severely deforms the real extent of areas, and is unsuitable for maps where the area should be presented as accurate as possible. Different projections give different perceptions of direction and space [Fig. 22]. For the area between 60 degrees North to the North Pole, we used Fig. 22 on the same scale, two different projections: a gnomone projection in the center of which is the North Pole, and Mercator's projection. The trajectory Kobenhaven – Anchorage seems quite different on the two maps, whereas in fact it is the same, only disturbed by Mercator's projection, as distances and directions are appreciated differently, according to the projection.

Topology teaches us to look at phenomena from a different approach than we have been educated in. A new approach gives us a possibility of having different

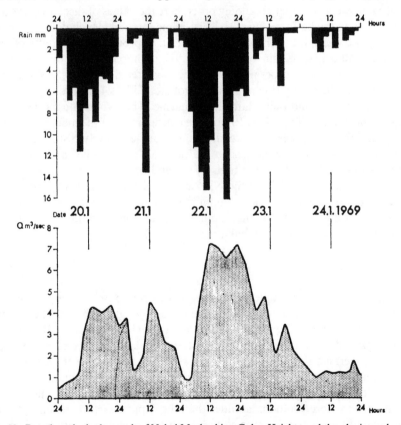

Fig. 23. Rotation: the hydrograph of Nahal Mashoshim, Golan Heights and the pluviograph of the area [Nir, 1974].

attitude to a problem, as it is illustrated it from a different standpoint and it enables us to compare and reevaluate. The major methods of topology, which can be used in regional geography, are rotation, translation and dilatation [Cole and King, 1968].

Rotation

We are used to considering and regarding phenomena in a certain order adapted by use and education. Maps today are presented with the North figuring at the top of the map; in the Middle Ages it was the East which figured at the top of maps, from which – Orient – came the word 'orientation'. We write from left to right in Latin alphabet, but from right to left in Arabic, Hebrew and Japanese. The abscissae of graphs are at the bottom of them, etc. Rotating from the conventional approaches can give us new insight into a phenomenon.

Let's say that our aim is to study the relationship between the rainfall and the discharge in the Nahal Mashoshim, in the Golan Heights [Inbar, 1972]. Instead of using the obvious method, where zero value is at the bottom of the graph and the accumulating values in ascending order, here the values of discharge are given from the top to the bottom [Fig. 23]. This simple rotation, which does not change anything in the graph but the way we regard it, improves the possibilty of comparing the influence of rain on discharge. The comparison of the juxtaposed pluviograph and the hydrograph reveals the close relationship between them.

A change in the orientation of a map by rotation [Fig. 24] can reveal a concept which otherwise would quite difficult to imagine. The two greatest rivers of the Mediterranean, the Nile and the Rhone, are considered, but the map of the Rhone is rotated by 180 degrees. Its direction is, of course, entirely distorted, but the rotation reveals many similarities to the position of the Nile: a shore-current coming from the left side of the both rivers, and bars on their right side; small anticlinal ranges to the right of the Nile, and the same to the left of the Rhone; ancient massifs on the right side of both rivers; and, of course, the delta in both cases. The rotation reveals that both rivers are situated in a similar geomorphological situation [Y. Itzhaki, pers. comm.].

Translation

Translation is a rather simple procedure, in which two phenomena under consideration, actually distant one from another, are brought together. The question under consideration is the extent of the quaternary glaciation in the Rockies and in the Alps, and the resultant differences in tourism, ski industry, etc. Among many factors concerning the glaciation of the two regions [distance from the ocean, altitude, exposition of the slopes, etc.] is their geographical situation, i.e. their latitude. The translation shows clearly [Fig. 25], that two thirds of the Rockies are situated in more southern latitudes than the Alps; therefore, all other factors but equal, to this factor of lower altitudes can be attributed a certain role in the smaller extent of glaciation in the Rockies.

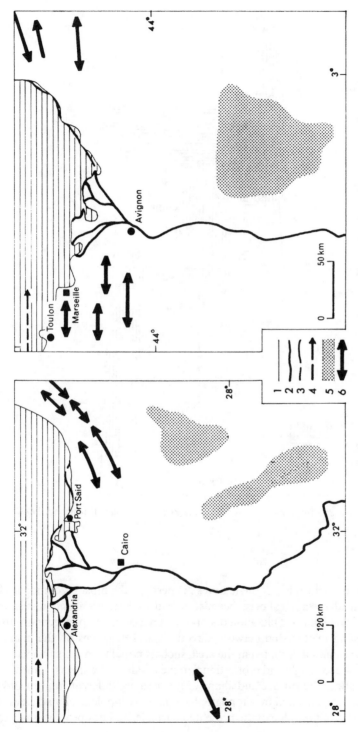

Fig. 24. Comparison of the Nil and Rhône deltas, by rotating the later. 1. Sea shore; 2. River; 3. Bar; 4. Sea current; 5. Ancient shield; 6. Folded mountains [Nir, 1974].

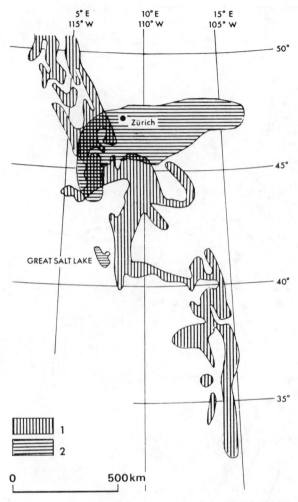

Fig. 25. Translation: the Alps and the Rocky Mountains [Richmond, 1969].

Dilatation

Dilatation, or 'deforming of the features', is perhaps the most revolutionary form of topological metamorphosis, because it causes alienation and distortion of topographical features. Dilatation distorts the area of the map by giving – instead of the existing relationship between directions and distances on the map – the values of variables distributed in the area, such as population density, income per capita, etc. Accordingly and proportional to these values, the area is shaped. When using dilatation, the surface and shape of the area are different from its physical properties as represented by a topographical map. A topological map gives areas proportionaly to the value of the presented variable, and not proportionaly to the

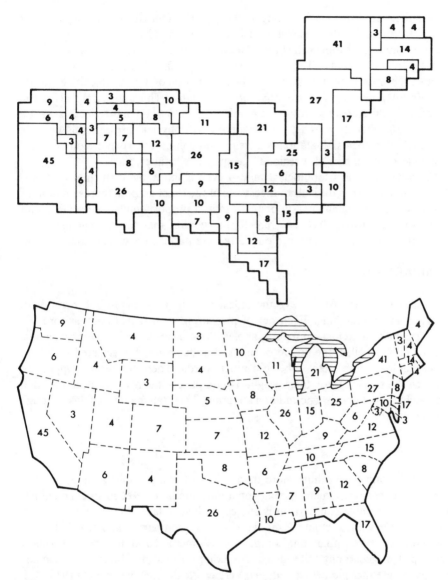

Fig. 26. Dilatation: the number of electors of the president of the U.S., represented on a topographical and on a topological map, 'Haaretz', November 6, 1972.

physical area of it. In Fig. 26 two maps represent the number of electors in the 1972 presidential elections in the US. The lower map is a topographical map of the different states, where the area on the map is proportional to the physical area of the states, without any relation to the number of the electors; this figure is writen down within the area of each state. In the upper map, the areas of the states are distorted because the map is not giving information on the area of the state but instead the area is made proportional to the number of the electors in the particular state. On this map, the greatest area is occupied by California, not by Texas, because California supplies the highest number of electors. The area of New York is also greater than that of Texas, and so is Pennsylvania. Texas is fourth, because the area represents the state's value in the distribution of the variable under study. *The area is proportional to the value of the variable.* The upper map gives the hierarchy of the states according the number of electors; the more electors a state sends, the greater its area is, proportionally. Its physical area, in this case, is neglected. Topological dilatation is a visual and comparative expression of socio-economic relationships in a regional, geographical framework.

The advantage of a topological map – even if the visual distortion of areas, shapes, boundaries is difficult to digest – is in the exact graphic representation of a certain variable, phenomenon or process in a proportionally true scale, as an object of comparison and research. Small wonder that topological maps are widely used in research and in didactic representations of the research results.

COMPARISON

Comparison seems to be one of the oldest and the most persistent methods of thinking and researching. Each act of measuring is in fact a comparison. We compare different items – tomatoes, apples, flour, etc. – by weighing them against a metal cube labelled one pound or one kilogram. When the weighted item is in equilibrium with the metal cube – in other words, when the visual comparison between them, as regards their weight, is complete – then we are declaring that the weight of that item is a pound or a kilogram. This statement is reached through comparison.*

Analogy

Logical reasoning is nothing more than a way of comparison; analogy is comparison between phenomena having apparently similar or even identical qualities. Therefore, a reliable comparison is subject to logical laws. We cannot compare just anything with anything else; we can only compare items which have a common basis for comparison. Comparison as a method of study was not specially developed in geography, although many comparisons in subject are are made intuitively. Hettner pioneered with his 'Vergleichende Laenderkunde' [1934–37], but comparison is not accomplished in each of his case-studies. In most cases he merely juxtaposes the compared items one against the other, whith the comparison actually left to the reader.

A logically based comparison can be a central method in regional geography. By comparing different regions we can classify them, evaluate their potential, or discover a certain phenomenon which makes a certain region different from others. The importance of a comparison is the lesson learned from it, which issues

---

* Today, of course, electronic weights do not show visually the act of comparison.

from the analogy between the compared subjects. Comparison should not be confused with identity; it is not argued that in a comparison all elements should be identical.

Analogy is the basis to each comparison. We know certain qualities of a variable, and some qualities of an another one. If that known part of the qualities of the second variable seems to be similar or even identical to the qualities of the first, we can suppose that perhaps other qualities as well of the first variable – which are unknown to us – may be similar to the parallel qualities of the second variable. By analogy, we assume the unknown qualities of the second variable. In a comparison, therefore, we must be sure, that at least some of the qualities of the compared variables [in our case – zones, regions, their parts or qualities] have a common denominator. We can consider *variables* [climate, degree of education, income per capita] in *places* and in *time*. These three elements – variable, place, time – are appropriate for creating different components of comparison, which not always are relevant. By the three elements – variable, place, time – there should be a common denominator of two of the compared elements: *one* variable in different places at *the same* time; or *one* variable in *one* place at different times; or different variables in *one* region at *one* time. It is impossible to compare different variables in different regions at different times [Cole and King, 1968].

'Pars pro toto'

Another type of inquiry is by comparing a part of a phenomenon to the whole phenomenon. By comparing a certain variable in a region with the distribution of that variable in the whole country, we can see whether in that region the variable diverges from the whole. Our case-study is as follows: is the season of cultivation of vegetables in the region of Bet Shean, which is a semi-arid region on the fringe of the desert, different from that of the Mediterranean parts of Israel? Our assumption is, that climatic constraints seem to impose seasonal differentiation from the Mediterranean part of the country:

Seasons of cultivation of vegetables in the year 1957/58 in Israel and in the Region of Beth Shean [Nir, 1968]

	[as % of the area of vegetables]			
	Summer	Automn	Winter	Spring
Israel	12	22	13	53
Region of Beth Shean	2	63	13	22

There is indeed a very great difference in the growing seasons of vegetables in the country as a whole and in the Region of Beth Shean in particular. Whereas the main season in the country is the spring, continuing into summer, in Beth Shean it is the automn, because the winter there is mild and vegetables planted in automn can survive, whereas the spring is too short and too hot.

COMPARATIVE REGIONAL GEOGRAPHY

Perhaps the most difficult comparison in geography is the regional comparison, a comparison between two regions. There is no reason to compare two regions, if there is no apparent possibility of analogy between them. On the other hand, it is very instructive to compare two regions which possess some common geographical elements. Certainly, the most instructive lesson can be drawn from regions whose populations face similar environmental challenges, and specifically from studies of the human answers to these challenges. I will take as an example a comparison between Provence in France, and Judea in Israel.

Both regions exhibit common landscape elements which we call 'challenges', as they demand an answer in order to transform the environment into an inhabited region. The terrain in both is built of limestone and dolomites of the Mesozoic period; the style of mountain-building is a simple folding, more intensive in Provence than in Judea. Karstic phenomena are common. The problem of water supply, an issue of their mediterranean climate, is acute in both regions. Storage of water, or supplying it from a distance, are the primordial problems and have been a part of the struggle for survival since the regions became eucumene. Autochtone water resources are sparse and poor: underground water-table in the plains, and vauclusian karstic springs in the mountains.

The analogy between the two regions, as related to the importance of water, is reflected quite persuasively in the biblical legends of digging and care of wells, and in the novels of Marcel Pagnol *Jean de Florette* and *Manon des sources* [1962]. The two regions experience strong winds, especially in springtime, which do harm by their drying effect on crops and trees; the 'sharav' or the 'khamsin' of Judea, which destroys the flowers of the spring with its high temperatures and dryness, and the 'mistral' in Provence, which destroys flowers by its velocity and low temperatures, sometimes linked with frost.

The traditional agricultural cultures are the same: wheat, vine, barley, olives. The proximity of the sea and their situation on crossroads between east and west, north and south, can also be seen as analogous. Even their historical development, which certainly was different, include periods of five hundred years when both regions were exposed to the same political and cultural influence of the Pax Romana. In both of them are remains of Roman highways, Roman aquaeducts, Roman baths, Roman water-mills. Even in their earlier periods, the Gallo-Roman 'valum' and the Cannanite-Hebrew 'tel' can be compared, as well as their Mediterranean acropolis-settlements.

The post-Roman periods brought new, dispersed, forms of settlement, such as the bastide and mass in Provence, and the khans, kasrs and deirs in Judea. The forms of 'genre de vie' on the fringes of the settled realm – the semi-desertic fringe of Eastern Judea, and the karstic parts of Haute Provence – are comparable up to the recent past: the transhumance of the sheep from the Rhone Valley and the Lower Provence to the summer pastures in the Haute Provence, and the semi-nomadic genre de vie of the Beduins in the eastern part of the Judean Mountains, which persisted until the 1940s [Shmueli, 1970].

Notwithstanding these apparent analogies, the differences between the two regions are perhaps just as great. We cannot compare the functional importance of the port of Ashdot [built in the sixties] or of the very ancient, and very small port of Jaffa to the importance of Marseille; nor is Manosque equal to Jerusalem. Only by studying the different development of the anthropogeographical elements can we understand the differences between the two regions. In the nineteenth century, they were entirely different: Provence was a section of a modern, developed and industrial nation, Judea was one of the backwaters of the declining Turkish Empire.

The twentieth century, and particularly since the 1960s, has seen a narrowing of the gap between the two regions. Industry, tourism and modern agriculture, developing in both regions, make the comparison between them more acute than at the beginning of the century.

The basis for a comparison between regions is their environmental challenges. The answers are made by the populations. Here lies the most important issue of comparison: why are the *differences* in the compared regions as they are? This stresses the importance of the input made by populations. By comparison, we can give appropriate weight to the contribution of a population to the lived region.

A regional comparison can be focused on a certain element of the compared regions and result in an establishment of a *model of development*, which can have a general value. The role of the comparison is to attribute to each region its place in the model. The study of Jackson and Sofer [1986] on the development of irrigation schemes in Utah [US], Israel and Jordan serves as a case study. The thesis of their study is that an examination of irrigation systems in these three countries leads to a hypothesis that there are marked similarities both in the stages of development of irrigation agriculture and in the use of the potential water resource.

The areas studied are the Utah Valley in the US, the Bet Shean Valley in Israel and the Eastern Jordan Valley, or the Ghor Irrigation Scheme in the Hashemite Kingdom of Jordan, which lies opposite it. After examining the different types of irrigation over the last hundred and thirty years, the authors propose that the irrigation system be seen in five developmental stages:

1. *Initial stage*, with simple diversion systems based on individual land ownership;
2. *Community organization stage*, with irrigation systems organized for the benefit of the community;
3. *Regional stage*, with large scale reclamation projects affecting whole valleys;
4. *Interbasin transfer stage*, with even larger and more marginal reclamation projects;
5. *Rationalization of water stage*, with emphasis on the use of existing water rather than the generation of new water supply [Jackson and Sofer, supra cit., p. 388/389].

Within this model of development, the actual development of each region is plotted on a time scale [Jackson and Sofer, 1986, p. 400]:

	Utah Valley 1849	Ghore Scheme	Beth Shean
1850			
	stage 1		
1865			
	stage 2		
1880		stages 1 & 2	
			1882
1895			
		stage 3	
1910			stage 1
1925			
			1930
1940			
			stage 2
			1948
1955	stage 4		stage 3
		1959	1959
		stage 3	stages 4 & 5
		1979	
		stages 4 & 5	
1984			

It seems that stages 1 to 3 – from the individual to the regional scheme – are reached at a relatively early level of activity; degree 4 was reached at a relatively late stage. Continued population growth in arid and semi-arid regions indicates that all irrigation societies should eventually move to stage 5, but only when there will is a national consensus on rational allocation of water resources; this stage still has not been attained – according to Jackson and Sofer – in Utah.

This method of comparison, using a model of development to compare common variables in different regions, can be an important tool in comparative regional geography.

CASE STUDY

The case study method is an important procedure in regional studies [Jean Gottmann, letter to the author, March 6, 1968]. It should be considered a method comparable to observation in biological studies, or to an 'individuum' in the medical sciences. The processes going on are but fractions of general processes. From an 'individuum' therefore we can learn not only about it, but also about the general processes conditioned by a discreet localisation.

An expanded type of case-study is the 'regional monographs', which studies a certain case *in extensio*. The nature and quality of regional monographs depend on the period in which they are produced; as statistical awareness was not basic to human geographers until the 1940s, regional monographs, culminating in the studies of 'pays', were mostly qualitatives and their essence was the study of the 'genres de vie'.

The main difference between the now classical regional monographs and the modern case study is mostly a matter of the extent of the investigation, although the approach to the theme is also different. Any subject in any realm can figure as a theme for a case study, as a which is *a detailed study of a phenomenon*. In regional geography, this phenomenon posseses a certain territorial and individual completeness; but the aim of the case study is to serve as an illustration to a study of a larger problem or region. A village in Malaysia [Clarkson, 1968] can be treated as a case study in tropical agriculture; a suburban neighbourhood can be a case study in social relationships; a section of a highway can be a case study in interurban communication system.

Perhaps the most important role of case studies in regional geography, following the concept of systemic region, is that they are the *individual bits of stone in a mosaic*, from which can be built the whole mosaic, as an entity. A classical example is Pierre Gourou's approach to region [1947, etc.], which is based on dozens, and indeed hundreds of case studies, from which he draws general opinions on the wet-hot tropics as a great regional entity where there exist interwoven relationships between diverse and numerous variables. This is a regional geography very different from enumeration of variables. Perhaps we can see in a case study a *natural model* [Chorley, 1962], on a real scale, of a certain reality which is to be studied by critical observation. The danger in the interpretation of a case-study is hasty deduction from a single case to a whole statistical population; but the significance of a case study can be evaluated, as in each model, by checking its representativeness concerning the population from which it has been drawn. Neglect or negation of this method, merely because a single case cannot be considered as a principle or a rule, is a negation of scientific research.

CONTRIBUTION OF REGIONAL GEOGRAPHY TO SOCIETY

Thus regionalism has what might be called an iconography as its foundation: each community has found for itself or was given an icon, a symbol slightly different from those cherished by its neighbors.

Jean Gottmann, 1962

APPLICATION OF REGIONAL GEOGRAPHY IN PLANNING

The application of a geographical regional study besides its intellectual contribution to a better understanding of the economic, cultural and social problems of a discreet region [Juillard, 1962] can be canalised to three main uses: to improve the cooperation of different regional elements by administrative regionalisation; to treat special regional problems; to use the regional study as the basis for planing [Mazúr, 1983].

Regional geography in administrative organisation

Considering administrative repartition of space as arbitrary, as compared to physical repartition which seems 'natural', is erroneous, especially in older countries of continuous civilisation, where the administrative boundaries created throughout history real-life facts and structures. Perhaps in a new, sparsely populated country, the administrative delimitations preceded human achievement, but even then they influenced and created new frameworks of activity, relations and habitat [p. 120]. The administrative framework is the vehicle of regionalization and of organization of public services, which develop within administrative boundaries and enhance local activities. On the other hand, administrative patterns created in the far past according to the criteria relevant to those days – means of transport, accessibility, communications, security, strategy – are today, in many cases, obsolete. Rather than a vehicle for development, the old administrative pattern are obstacle to it [administrative units too small for economic or other operations, for example, or the inadaptability of existing frameworks to manage new phenomena emerging in the region].

Changes in the administrative organization can be illustrated by a comparison between the size of administrative units in the past and the tendencies existing today. If we take as the smallest administrative unit the area of a commune or a small town, which in older developed countries are between thousand and fifteen hundred hectars [10 to 15 sq.km., Nir, 1974], then the next administrative unit, a canton or a sub-district the area of which is between 150 and 300 sq.km., and which was a meaningful organic unit in the technological situation of the past, is today of very limited administrative value, except for electoral and fiscal purposes. As means of transportations and communications developed, so developed and grew the administrative units. The revolution in transportations by the railway in the first half of the 19th century in Europe and in the eastern US, and later in that century across the world, created greater administrative units, having areas of thousands of square kilometers [Regierungsbezirk in Germany, Departement in France, etc.]. In this century, even larger administrative and political units – which became, as we mentioned above, also a framework for cultural, economic and social activities – have been created: the German 'Land' has areas as large as 25.000 thousand sq.km.; Italy was divided into 19 basic units [1948], England into 11, each of them of fourteen to sixteen thousand sq.km.; France was organized in 1955 into 21 regions of activity [Hansen,1968], as compared to the repartition into 92 departements. This reorganization into such large units, espe-cially in France, is concentrated on carring out regional projects [p. 153]. The new administrative delimitation and reorganization is based on lessons taught by regional geography.

The opening of administrative frameworks exists also on lower levels of administrative organization: cities are amalgamated into conurbanisations, ham-lets into rural communities. 'Metropolitan areas' replace older, obsolete adminis-trative units which are too small to fulfil their original purposes.

The changes in the size of administrative units in most of the countries over recent decades, in both the old and developed as well in the new and developing countries, are that administrative frameworks, although not easy being changed, must have nevertheless adjust themselves to the facts of life, in order to be efficient. Geographical regional research has an important role in backing these changes.

Regional geography in relation to regional problems

Admistrative units, once delimited, serve as a relatively permanent framework of activity. But some functional regions, dealing with a certain problem, are created ad hoc and their delimitation later serves as the executive framework for soluting the problem [Juillard, supra cit.] Three groups of problems related to regional geography can be discerned:

- Problems related to the relationship man/environment: flood control; control of coastal erosion; draining of swamps; struggle against desertification; wildlife reserves, etc. Relating to these problems required an ad-hoc delimitation of the problem-area, which is not necessity congruent with an existing administrative unit. Its delimitation will be aided, if nct imposed, by regional geographical research.
- Problems related to the need for economic development: reforestation of badlands [Spain]; development of 'virgin soils' [Khazakhastan]; irrigation projects [California, Pakistan].
- Problems related to socio-economic challenges: development of a region struck

by poverty, lack of resources, unemployment [Appalachia, [p. 9]; a region of depopulation; a redevelopment of an urban center, etc. These problem-regions can be converted, within their ad hoc limits, into planning regions, or reorganization regions. The remedy proposed will be useful only if based on a regional basis, which can reveal its basic problems. This group of problem-regions leads us to an integrative research, i.e. to the use of regional geography in the overall planning of a region.

Regional geography in planning

Regional planning on a world-wide scale in the post-war years enhanced the development of the regional method of study [Lavrov & Sdasyvk, 1984]. One of the qualities of regional planning should be an overall and integrative consideration of the region and its problems, an approach close to our concept of the systemic region. The best way to ensure good planning is team work, including the contributions of architects, economists, sociologists and geographers to the common goal. Excellent planning made by an architect may be doomed to failure, if sociological research was neglected, or if the physical challenges – climate, soil, hydrography or distances – were not taken into account [Glikson, 1955].

Although the recent decades have seen intensive planning on local, national and international levels, planning is not new to mankind. It was widely used in the past, especially in centralist monarchies. The differentiation between regions is, generally speaking, the result of lengthy historical development; a region is a result of human activity. To mould geographical material is the main activity of each administration [Gottmann, 1966], as no political action can be indifferent to regional differentiation. The alternative of politics was either to preserve the regional differences of a political unit, or to try to modify them. The first alternative was always wiser, as an attempt to change regional character usually meets the antagonism of the region's population. But even if the will exists to preserve the existing differences in a certain region, the daily routine and decisions of the central administration of the nation as whole also influences the particular region in its daily activities. Therefore, each administrative activity should be effected on two levels: on the level of the central administration, which decides on a certain policy, and on a regional level, which leaves to the local regional authorities the execution of that policy. In our century, regional planning is a routine part of the administrative organisation of many nations.

For purposes of planning, the region may be considered as a 'means for action' [Dziewonski, 1967]. The delimitation will be defined a priori, following the planning aims, or according to a preliminary regional research which imposed the boundaries of the region within which should be act. *To act means to change the actual state of things*, according to economic or political aims.

Geographical challenges, including socio-economic challenges, are specific to each region. The planner may wish to propose a pattern of life in occupied realms, or change patterns of life of an existing population. In any case he must carry out empirical research on the landscape in which he wishes to act [Glikson, 1955].

This intimate knowledge will serve him as a background to his planning, and will progress his work with relevance to the conditions of existence of the people living in the area. Planning should always be carried out by methods which suit the phenomena that characterize the distinct region.

One of the basic paradigms of regional geography, which can be of great help to a planner is that the perception of environmental challenges differs among different ethnic groups. The potential of a region is not a given fact, outwardly perceptible as such. As percepted by different groups of population, the same element may or may not have a certain value for the society. Therefore, *planning should be oriented towards the particular society*. The planner should know for which population, for what social, cultural and ethnic perceptions, the plan is designed. In other words, the goal of the plan determines the character of the planning. The planner's task is not an easy one. The aim of planning is to maintain or achieve an equilibrium between the environment and the social group that occupies or will occupy the region; even if the environment can be considered as more or less constant, this is not true of social groups which are open to perpetual change.

Most planners are convinced [Glikson, 1955] that the preliminary research of the basic phenomena in a region is a schematic and boring part of their art, and that the decisive part of planning is the applicating their creative immagination, conceived as the most important quality of gifted planner. Of course, imagination is a very important component of the planner's personality, but a true basic, empirical research should be much more than a mere documentation of facts, an inventory of the region. It should be a mental adjustment and practical preparation to the act of comprehensive planning. This means research deeping into the problems of the population at which the planning is aimed, and into the environmental challenges, completed by an effort to integrate the two. From this standpoint, basic research aims of the planner are identical to the aims of the regional geographer.

Two examples of regional planning

I wish to illustrate the use of regional geography in planning by two examples, which are today perhaps classic ones: regional planning on a national level in France, and the planning of the central Israel region of Lakhish [Bendavid, 1970]. Examples could be brought, of course, from almost any country in the world, especially the socialist countries where planning is an integral part of domestic policy.

Although the regional planning in France is one of the most advanced in the so-called Western world, *regional* planning there did not always enjoy a national preference [Gottmann, 1958]. A centralist tradition and a population which enjoyed a certain stability, were not favorable to regional initiatives and provided no background for great modernistic enterprises [Dickinson, 1947]. But, as in almost all parts of the world, in the aftermath of the Second World War the need for a new organization of the nation was urgent enough. The growth of population and the imperative of absorbing one million French settlers from Algeria, together with perception of a need for regionalism and regional policy [p. 156], all ripened into the decision for a comprehensive regional planning of the nation.

At the root of the perception securing nationwide planning was awareness of the growth of Paris and its suburbs at the expense of other parts of the nation. In the last hundred years Paris had absorbed the major part of resources for development and growth. Plans for decentralisation, far away from Paris, urged a new reorganization of industry. It was assumed that the decentralization could be achieved only by recovering backward regions where the economy and demography were in decline. On the other hand, some regions needed planning not only for economic but also political reasons, as the depriming economic situation revived old separatists sentiments of the Bretons, Basques, etc.

Until 1955, each ministry executed its activity in the different *départements* according to its policy, unrelated to the policy of other ministries, by the intermediary of the *préfets*. Therefore some principal cities, residences of the *préfets*, were dealing with the activity of twenty to thirty subdistricts [Hansen, 1968]. On the other hand, the *préfets* were dependent on their ministers, and the main decisions were made in Paris.

Fig. 27. Regionalization of France: the 'départements' and the planning regions [Hansen, 1968].

In June 1955, the government declared the 'Plan d'action régionale', with the goal of stimulating regional economy, especially in regions suffering from high unemployment or un-satisfactory economic development. The plan had to coordinate the activity of the different administrative institutions with local public institutions. The framework was established in 1956, when France was divided into 22 planning regions, comprising the 94 existing *départements*. This division was later altered, and today there are 92 departements in 21 planning regions [Fig. 24].

This nationwide regional planning was based on studies that revealed the striking socio-economic inequalities between the different regions of France in the 1950s. I shall illustrate them by some parameters. In 1965 the percentage of employed in agriculture in the region of Limoge was [Hansen, 1968] 27.5% of the active population, and in the Nord only 3.5%. The case of Limoge appeared at first glance as exceptional, as the second region after it, Bretagne, had only 18.7% employed in agriculture, close to Normandie with 17%; all other regions had less than 15%.

As for secondary education, Limoge was at the bottom of the list, having the rank of 21 out of the 21 regions. The negative correlation between the number of those working in agriculture and the level of secondary education was quite evident. Even a superficial consideration of the geographical facts reveals the inequality between the different regions. The importance of regional geographical research as the first step to regional planning is obvious.

Lack of relevant data is one of the handicaps of regional planning; another is the lack of cooperation by small local units, affraid of being damaged within a larger, regional concept. To advance the realization of the Plan, a new institution was created in 1963, representing the population in planning, 'Delegation à l'aménagement du territoire et l'action régionale', or DATAR. This did not evidence intention to build a new administrative framework to rival the existing administration, its purpose was advance initiative and regional cooperation.

Example of planning on a regional scale is that of the region of Lakhish, in the center of Israel, some seventy km south-east of Tel Aviv. This was the first region in Israel organized and settled according to a regional planning [Bendavid, 1970].The work began in 1956. A team of architectcs, economists, agronomists, sociologists and geographers lived for two years in the region and prepared a plan, consisting of the research of the physical background, agricultural organization, regional economy, urban planning, agricultural planning, and sociological and demographical research of the population to be settled in the region. Thirty four maps analysed the topographical, climatological, and pedological aspects, proposals for land use, transportation, administration, education, health services and cultural services. The theoretical model was that of the 'central place' of Christaller: three to four settlements concentrated near a rural center [Fig. 28]. Four such nuclea surrounded a brand new town, Qiryat Gat. Unlike the planning in France, here the planning was based on 'tabula rasa'. From a perspective of thirty years, the planning was an economic and social achievement and later served in developing the scheme of Taanach, in the Valley of Esdrelon.

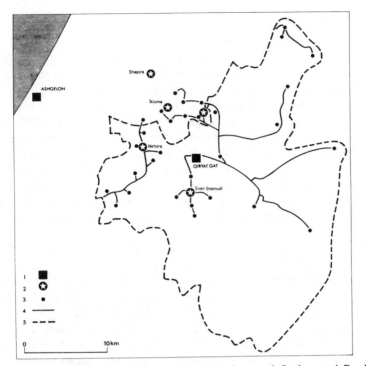

Fig. 28. Planing the Lakhish Region, Israel: 1. Town; 2. Rural center; 3. Settlement; 4. Road; Limits of the planning area [Bendavid, 1970].

REGIONALIZATION, REGIONALISM AND REGIONAL DEVELOPMENT

Regionalization and regionalism are two different concepts. Regionalization is the application of regional geography in planning and in development of the sub-divisions of nations:

a necessary procedure, which distinguishes regions which are characterized by one or more qualities [Zonneveld, 1983].

'Regionalism' first appeared in 1874 [Gilbert, 1960a, 1960b], and since 1890 has been commonly used to mean the idea of political organization on a regional basis [Mackintosh, 1968]. Today, the meaning of *regionalism is the mental perception of a region*, a sense of belonging to a region. It has a distinct political flavor. Of course, overlapping, influences and cross-implications exist between regionalization and regionalism. Regionalization can be seen as the framework, and regionalism the content of it [Labasse, 1955].

Regionalization

Regionalization is sometimes perceived as classification, [Bratzel and Muller, 1979], a repartition of the earth's surface into discreet units [Spence, Nigel and Taylor, 1970]. But classification is not proof of objectivity. The inherent nature of a problem means that classification remains subjective.

Even the most centralised states are divided into sub-divisions, situation which dates from the dawn of history [Gilbert, 1960]. A given regional political unit – the state or nation – is not necessarily the right size for economic development to benefit those whose need is greatest [Schumacher, 1975]. It is generally too large, and develops into core and peripheries. The result of 'development' can be a fortunate minority as compared to a majority left even poorer than before. *If the purpose of development is to bring aid to those who need it most*, each 'region' or 'district' within a country needs its own development. This is the political basis for regionalization.

Regionalization is a spearhead in both applied and pure geography [House, 1970]. Regions are studied with a view to more effective order in economy and society. Regionalization also studies the regional impact of governmental policies. In many cases, it can be motivated by the state. Charles de Gaulle promoted regionalization in his presidential campaign of the late 1960s.

Economic regionalization is part of the domestic policy practicied in the so-called Eastern Bloc. In the USSR, regions are considered living realities [Polkshishevskij, 1975]. This concept is generally accepted through the breakdown of the USSR into territorial complexes. The lower level of the taxonomic scale are industrial centers and agricultural regions. They became larger units, not all having the same taxonomic interpretation, but their objective existence is not denied. Then come still larger units, whose delimitation is not universally agreed, but the reality of which is not doubted. Thus, the regionalization process is neither a dissection of the country into territorial parts nor a classification for convenient analysis, but a *systematic uncovering of actually existing integrated territorial units*.

Regionalization has a focal role in Soviet economic geography. It is the emerging outlines of the future development of a given territory, with regional planning looking ahead twenty to thirty years. The term 'region' is used in the USSR as a concept steered by economic considerations, delimited in most cases by administrative boundaries. This concept of region is not that of the systemic region.

Regionalization is the true application of regional geography [Le Berre, 1980], as the goal of regional geography is the management of territory. Rather than tension feeded by regionalism between different ethnic groups and the central government, regionalization can moderate the political, social and economic differences between the region and the central government. Regionalization cannot, however, diminish ethnic and cultural ambitions. In comprehensive regional planning, a niche should be found for these ambitions as well. The adaptation of this approach in Catalunya, f.e., where the Catalonian and Spanish languages are both official, shows the success of comprehensive regionalization.

Regionalism

Regionalism is both a collective feeling and a individual's attachment to the region in which one lives. In many cases, it wears the dress of political movement among minorities which claim the right to be different [Le Berre, 1980]. In 1898, the 'Union régionaliste Bretonne' was founded in France [Dickinson, 1947], and in 1900 the 'Federation régionaliste Française'. The aims of these organizations were to decentralize administration, economy and social activities from the capital.

As distinct from regionalization, *regionalism is a mental state* indicating a perception of attachment to a certain region [Knight, 1982]. This perception can have either positive or negative attitudes for the larger framework of which the region is a part: Hamlin speaks of 'nordism', an attachment to the landscape of northern Canada. The regionalism of the Basques in northern Spain and southern France developed into a political separatist movement.

On the other hand,the perception of belonging to a place leads to topophilia, love of place [Tuan, 1976; Wright, 1966b; Pivetau, 1969] and to other expressions of emotional links between people and region, now praised and studied by the humanistic trend in geography. There are also poets and novelists who have a deep regional feeling [p. 72]. In Britain, great stimulus to the development of region-alism between the two wars was provided by broadcasting [Gilbert, 1960], in the establishment of regional broadcasting stations that contributed to a regional consciousness. The regional television network after the war also contributed to this trend. In Britain roots exist, from which regionalism can be fed [Philbrick, 1957, 1960]. These regionalistic tendencies clash with the spreading of global of uniform culture [Nir, 1985]. Regionalism is one of the ways preserving, and even supporting regional and local peculiarities and individuality. The goal of region-alism is to move interest toward ethnic or social groups which are different from those of the national collectivity.

Regional development

Regionalization and regionalism meet in regional development [Mierny, 1982], which can stem either from a political plan issued by a central government aiming to achieve political and social gains on a national scale [France for the develop-ment of Auvergne, Bretagne, Languedoc, Spain for Catalunya and Vizcaya, Brazil for the Nord-Este], or from a movement from below, from the regions in question, due to a strong regionalistic feeling – or by a coordination of both [Sanguin, 1984]. In this perspective three interpretations of regional development are discernible [Kuklinski, 1985]:

- *regional development as a fact of the objective reality*, conceived as the develop-ment of regions selected and delimited according to indicators based on reliable statistical data and comparative analysis. This type of approach believes in quantitative, value-free research approach.
- *regional development as a socio-cultural and political movement*, where the driving force for it is pressure from below, fostered by socio-cultural and political

dissatisfaction growing out of what we defined above as regionalistic feelings. According to this interpretation regional development is a socio-cultural or/and political movement, based on an ideology having both rational and irrational elements.
- *regional development as an activity defending the political status quo*. Here, regional development is a part of the program of the authorities, adapting a national policy of controlled change, where the development can be interpreted either as an autonomous goal for itself, or as an instrument for achieving economic, political, social or military goals. In both cases, regional development is used for executing national policies.

The future of regional development depends on the interaction of the forces coming from the region below and coming from the central authorities.

FREEDOM AND GEOGRAPHICAL ACTING

Elements of a cultural landscape [p. 57] – settlements, roads, fields, factories, parks and gardens – cannot be explained without the relating them to the freedom to act, to move, to build, to plant. This freedom of acting can be enhanced or restrained by technological, economic, political and socio-cultural factors.

Freedom and technology

Technological improvement of means of production had immense influence on the organization of labor and by that on the social and political structures of humanity [Gottmann, 1981b]. The abolition of slavery in Eastern Europe and in the US in the middle of the nineteenth century was a consequence not only of minds enlightened by the ideas of Jean Jacques Rousseau and the French Revolution, but also of the development of sources of energy other then the man's muscles: a steam, and later, electricity and the combustion engine.

Freedom of movement has gained enormously by the possibilities created by technological development of means of transports in the last hundred and fifty years. The technology of flight makes possible movements of man and goods to great distances in short of time and with increasing comfort. Never have such masses of people had the oportunity to travel for work or leisure as in recent decades; in the months of July and August more people criscross Europe for leisure and tourism than all the masses of people involved in the 'Wandering of Peoples' at the beginning of the Middle Ages.

Over the last two hundred years, industrialization, mechanization, automation and other forms of rationalization of human labor have considerably increased the individuals' freedom of action individuals, especially in highly developed economies [Gottmann, 1981b]. Our behavior today is entirely different due to the availability of time to use free as we decide, and without having to devote it to acquiring of income, preparing food, to drawing water, supply heat etc., as it was for the preceding millenia.

Political and social freedom

The techological freedom is only an option of freedom,as man is not free from economic restraints; so, real freedom should be also economic, and, not less important, political, cultural, religious and individual. The definition of freedom is, of course a very difficult one; it occupies the mind of mankind since the beginning of its existence. There always existed constraints which limit freedom, especially the individual freedom. We already [p. 3] proposed a definition of it: *freedom is the right to be different from other, but with no right to harm the other*. This definition is broad enough to allow more refined interpretation of freedom, but, on the other hand, it says the essential:in a society, an individual has the right to be himself, different from the individua that *together with him* constitute the society; but the individuum restricts himself not to harm the freedom of the others. In a world of nations, ethnical entities, religious believings, political opinions, each one has its right to exist on the above stated condition.

Can absence – or avaibility – of freedom of an individual, of political freedom, or of some kind of totalitarian thinking be percepted in the landscape elements? To answer on that, we have to elucidate the problem of limitation of freedom.

Limitation of freedom

Even while advocating the freedom of the individual, it is clear that as man is a social being, living in societies, certain limitation of the freedom of the individual is required precisely to avoid causing harm to the freedom of others. This limitation, however, should not be imposed but made by concensus, by decision of the society, i.e by *laws*.

Each law limits the absolute freedom of the individual, but on the other hand, it reassures the freedom of the individual from the mishandlings of other individua.

In the individual's *use of space* some limitations to the individual's freedom to act have been generally recognized and accepted [Gottmann, 1981b], as trafic controls and regulations, planning of urban and rural spaces, border regulations, customs etc.

People restrict their rights quite rationally in issues as urban planning, regionalization of a city into business, residential, industrial and recreational quarter, even if the planning is in fact not other but a limitation of their freedom to built on the plots, which are their private property, what they wish and how they wish. Freedom is not unlimited; it is accorded to the needs and interest of the society, of which the individum is a part. Only by accepting a certain limitation to the absolute freedom of the individual, an *organization of space* is possible; but there are limitations stemming not out of a concensus of the population living in the region.

Great, imposant buildings, large squares where multitudes can be gathered, are appreciated by some architects as 'fascist' or at least as a centralist architecture. The architectural difference between the nineteenth century Paris and London is not issued by hazard.

It would be erronous, however, to attribute to ressemblant landscape elements to a reflection of political totalitarism, which imposes to the landscape a uniform aspect, following an ideology of uniformity. Architectural tendences of a government are not always a direct issue of a political tendency; in many cases, economic constranints of it are the deciding ones. Therefore, if we see in a landscape a certain uniformity, it can be the result either of a central factor or a free decision. Personal iconographies which stemm from a commun 'Weltanschaung' of the architects, or a common technology, can have the same effect:

> There are landscapes in America, separated by hundreds of miles, which ressemble one another in a striking degree, and many American towns, even many American cities, are all but indistinguishable as to layout, morphology and architecture [Jackson, 1972].

In many, if not in all Israel 'development towns' constructed in the 50s and 60s, the ressemblance and uniformity of the architectural style were not issued by a free decision of the population; all of the developing towns were planned, built and managed by governmental agencies. The issues is an urban landscape imposed by a central factor, not a free decision of the population.The landscape elements of a place, therefore, not always are the expression of the freedom of decision of its inhabitants. Nowadays, when private and not governmental iniciatives are constructing in these towns, the free decision of the inhabitants creates a new urban landscape.

Freedom means also a participation on the planning, as the use is that each planning of a neighborhood is subject to public criticism, where each one living in the neighborhood can interven by his objections. Freedom of an individual means also the choice of employment, the choice of habitat, the choice of shopping, the choice of leisure. In fact, each action of man which is a geographical one – and all the actions mentioned above are geographical – depends on the basic fact of the existence of freedom of an individual to make his choice. If this freedom is limited, than the geographical facts, resulting of this limitation, will be biased.

In extreme cases, the freedom of choice is one of the most important elements in the growht of the large cities in nations having great rural populations, as Mexico, Brazil, Peru, to mention only the most conspicuous cases. Instead of the freedom of choice, the pressure of their economic environment brings daily thousands of the rural poor population towards the great city, creating a problem of unplanned growht of the city, a 'metamorphosis of the metropolis' [Gottmann, 1981b]. We have to understand, that this rural exodus is rather a fact of *no choice*, as the economic situation eliminates the possibility of a free choice. The question is, if this 'freedom of no choice' could not be regulated by a governmental action, which would looking ahead to the future of the nation and try to influence its development by translating it into a true freedom of choice by a comprehensive planning.

Cultural freedom

Geographical phenomena, such as distribution of use of a certain minority language, are the result of a degree of freedom. So, f.e., the Welsh language in Britain is spoken today in a larger territory than a generation ago [Buchanan, 1977] as the result of a positive policy of Her Majesty's government and the choice and will of the Welsh people. On the other hand, the Basque language in Spain loses ground from decade to decade. If at the beginning of this century Pamplona was a Basque-speaking city, today the Basque languge is no longer spoken there. It was governmental policy neither to stimulate the expansion nor to preserve the Basque language. Poor schools for minorities are a known device for assimilating them into the majority. The opportunity to learn, to have access to information are but an expression of the degree of freedom of a population. In the end, they will be expressed in the landscape.

The opportunity to free research and to free access to the research of others depending on the degree of political freedom and unhindered free relations between nations. The impact of political systems on geography was studied by Lichtenberger [1979] in a case study of the differences between geographical research in the German Federated Republic and the German Democratic Republic:

Differences in Geographical Research in DFR and DDR

ASPECTS	DFR	DDR
Sphere of influence	USA	USSR
Organization	federal	centralist [Ac. of Scie.]
Guidance of research	individual, by Geogr. Res. Foundation	dependant on the Acad. of Science
Research perspective	diversified	concentrated at a few central projects
Regional foci of research	reasearch abroad dominant	no research abroad
Applied research	free market	Public commissions of the Acad. of Sciences
Publications	competition of journals	few journals supervised by Ac. o. Sc

Due to the different political approaches, the paths of research in the two neighboring countries lead in quite different directions.

There is no creation without freedom. Landscape elements are the result [p. 94] of mental elements of the people living in that landscape, metamorphised by the regulators acting on these mental elements. The degree of freedom afforded by the economic and political regulators acting either as constraints or as stimulants has a decisive role in the resulting landscape elements.

COMMITMENTS AND FUTURE OF SYSTEMIC REGIONAL GEOGRAPHY

La société se crée en créant son espace – les deux forment un tout indissociable qui entraîne la même évolution. [The society creates itself by creating its space; the two produce an indissociable entity which shares the same evolution (transl. D.N.).]

H. Isnard, 1978

Geographers – at least some of them – as well as people outside geography, are asking if there is besides its epistemological contribution, a socio-cultural message of geography. This is to say that geography should be some responsible, have some value [Fien, 1981] making geography worthy of existence as a social asset [Cowie, 1978]. This is, of course, a very subjective approach. As with most intellectual activities, geography can be used and misused [p. 19]. Morrill [1984] sees many facets of geographical responsibility, which can overlap and converge, but which also can conflict with one another. The first responsibility is to truth and scientific integrity. Science is the principal means by which a common understanding of knowledge can be advanced. The responsibility goes further is the discipline of geography: a geographer should have a rigorous training in both geographical content and geographical method. The geographer is also responsible for the teaching of geography in schools. The geographer's responsibility to the Earth results from his subject within humanist-environmental issues, followed by his responsibility to community, to society, to humanity. These last can create a conflict, or at least a dilemma: responsibility to any nation is, at least for some geographers, of primordial importance, and certainly can conflict with other values. Consider the Geographer working in his nation's War Intelligence Service [Bobek, quoted by Buttimer, 1983]: is he not in conflict over his dual responsibility to Humanity and to his nation? This conflict only can be resolved by the scholar himself, a decision of conscience which, of course, will be appreciated and judged according to the conscience of others.

Is there a place for 'values' in socio-cultural sciences in general, and in geography in particular? In positivist philosophy there is a tendency to separate 'values' from 'facts' and to analyse them 'objectively'; this leads to a moral impasse [Buttimer, 1974]:

> ... the 'objectivists' lead to a moral absolutism and dogmatism; the 'subjectivists' lead to a 'quagmire' of unique and uncomparable standards.

Neither approach deals adequately with the individual phenomenon. But there is a need to *evaluate* phenomena by certain aesthetic, moral and pragmatic criteria.

According to Buttimer, values may be considered on three levels: private, personal values [man as himself]; public values [man as a part of society] and professional values [vocation or discipline in which man acts]. In this regard, Buttimer understands values as a kind of *scientific humanism*: science is the guide for improving the quality of life and bettering the human condition.

Today civilisations – or perhaps better 'genres de vie' [Sorre, 1948] – in every part of the globe are in a permanent confrontation with a general and uniform civilisation, with a 'modern' genre de vie, erroneously identified with the use of technological innovations. Such a meeting of the local or regional with the global genres de vie is not a new pheonenon. It has always existed [Nir, 1985]. The Roman world influenced the greater part of Europe and the African and Asiatic shores of the Mediterranean, for several hundreds of years; the Chinese world inspired southern and eastern Asia, and so on.

In a uniform, standard world, region's pecularities are lost [Sorre, 1988]. Certainly, some pecularities are not to be regretted in their passing; but other elements included within the term 'genre de vie' are important to man's feeling at home in the region where he lives, to him not feeling displaced in his home.

Characteristic of our time are the problems of the *rapidity* with which innovations pass from one end of the world to the other, and the *cultural dimension* of a global civilisation [Mikesell, 1978]. At stake are not only the scientific and technological contributions, not only the exchange of raw material and finished goods, but also and principally uniformity of 'genres de vie' concerning clothing, popular music, literature, journalism, visual arts and communications [Gottmann, 1980]. It becomes more and more difficult to remain one self, the individual personality seems more and more submerged, in the whole. One might even ask seriously, wheter individuality in both material and spiritual aspect, is still a value or only a myth?

To this meeting or confrontation of civilisations different ethnic groups give different answers. In the great cities of western Europe we can observe a nearly complete disappearance of local or regional civilisations, particularly in their cultural appearances. Local characteristics have virtualy ceased to exist, in the style of urban centers and in the genre de vie of populations. Only the language of street signs signal the locating of a city in a certain country – and even this is not always the case, for the names of the great international companies are identical everywhere. The laws of urban structure seem already to have conquer the local and regional peculiarities of the central urban areas.

On the opposite extreme are those who totaly reject the modern, uniform, global civilisation, something difficult to imagine as it means a saying no to electricity, motor-vehicles, the telephone, as well as, of course, to the communications media of such as television, broadcasting, movies, and international magazins. Nevertheless groups do exist, most of them religious, which are ready to renounce this global civilisation. The most well-known are perhaps there rural Amish and Mennonite communities in Pennsylvania. Today they practicize nearly the same 'genre de vie' as their forefathers did two hundred and fifty years ago. Another example, this time of an urban population, is the case of the 'Neturey Karta'

['Keepers of the city'], a community of several thousand people in Jerusalem. They have their own legal system and do not respect national or municipal laws, and they boycott television and all mass media, they do not serving in the army. Their genre de vie is entirely detached from that of their neighbors, although they do use, in a very sophisticated way, some technological innovations, as well as modern hospitals.

Between these two extreme positions, exists a broad spectrum of possible solutions to the meeting or clash between the local and the uniform civilisations. One of the best examples here of succesful cooperation is the case of Japan [Gottmann, 1981a].

This country, third in the world in level of industrial production and having one of the world's most efficient infrastructures of public transportation, is in the front of the modern, scientific nations. Nevertheless, it did not renounce on its spiritual values, on a certain local genre de vie, its iconography. The japanese civilisation and the uniform, global civilisation have coexisted one with the other for more than a hundred years now, and the conflicts between them seem to be very limited.

Each of answer given to the meeting of the local and the uniform civilisations is legitimate if it results from a free decision of the local population. The lesson of the systemic regional geography is that man has the right to be himself, to live his own genre de vie, free to decide between the local and uniform civilizations. Regional geography becomes a vehicle for understanding and honoring the different genres de vie.

Systemic regional geography has a very important social role in our time of uniformity struggling with pluralism. Our world is at a very important crossroad to its organization and nature: whether to move towards increasing uniformity, towards a global culture and uniform civilization, or to continue a pluralistic way of life with *unity on the basis of diversity*. The choice is between totalitarism and pluralism on the cultural and economic levels. It is here that the question of freedom – the right to be different from one another – should be posed. It is here that systemic regional geography, dealing with geographical spatial individua, is important. Systemic regional geography stressing the individual and the characteristics of a region, opens the mind to a perception of freedom, of the right to be different and of the right to an individual genre de vie.

BIBLIOGRAPHY

Ackerman, E.A., 1945: Geographical Training, Wartime Research and Immediate Professional Objectives, *A.A.A.G.* 35, pp. 121/143.

Ackerman, E.A., 1953: Regional Research – Emerging Concepts and Techniques in the Field of Geography, *Econ. Geogr.*, 29, pp. 192/197.

Ackerman, E.A., 1963: Where is a Research Frontier? *A.A.A.G.* 53, pp. 429/440.

Agnew, C.T., 1984: Checkland's Soft System Analysis – a Methodology for Geographers?, *Area*, pp. 167/174.

Aiken, C.S., 1977: Faulkner's Yoknapatawhpa County: Geographical Facts into Fiction, *Geogr. Rev.*, pp. 1/21.

Aiken, C.S., 1981: A Geographical Approach to W. Faulkner's 'The Bear', *Geogr. Rev.*, pp. 446/459.

Amedeo, D. & R. Golledge, 1980: *An Introduction to Scientific Reasoning in Geography*, New York.

Andrews, H.F., 1984: The Durkheimians and Human Geography; some Contextual Problems in the Sociology of Knowledge, *Transac. Inst. Brit. Geogr.*, New Series 9, pp. 315/336.

Anuchin, V.A., 1973: Theory of Geography, *Directions in Geography*, ed. R.J. Chorley, London, pp. 43/63.

Anuchin, V.A., 1977: *Theoretical Problems of Geography*, Columbus.

Bachi, R., 1957: Statistical Analysis of Geographical Series, *Bull. Inst. Intern. de Statist.*, 36, pp. 229/240.

Bailly, A. & J.B. Racine, 1978: Les géographes, ont-ils jamais trouvé le nord? *L'espace géogr.*, pp. 5/14.

Balchin, W.G.V. [ed.], 1970: *Geography, an Outline for the Intending Student*, London.

Barnes, T. & M. Curry, 1983: Towards a Contextualist Approach to Geographical Knowledge, *Transac. Inst. Brit. Geogr.*, New Ser. 8, pp. 467/482.

Barrows, H.B., 1923: Geography as Human Ecology, *A.A.A.G.* 13, pp. 1/14

Bartels, D., 1973: Between Theory and Metatheory, *Directions in Geography*, ed. R.J. Chorley, London, pp. 23/44.

Bartels, D., 1982: Geography: Paradigmatic Change or Functional Recovery? A View from Western Germany, *A Search for Common Ground*, eds. P. Gould & G. Olsson, pp. 24/33.

Beaujeau-Garnier, J., 1972: La géographie et la notion de systèmes, *La pensée géographique française contemporaine*, Mélanges offerts à M.A. Meynier, St. Brieuc, pp. 107/114.

Beujeau-Garnier, J., R. Brunet, P. Claval, F. Damette, A. Froment, Y. Lacoste, A. Reynaud, M. Blodeau, 1979: Vous avez dit région, *Espaces-Temps*, Nos. 10-11, pp. 10/41.

Bendavid, A. [ed.], 1970: *The Lakhish Region: Background Study for Research in Regional Development Planning*, Settl. Study Center, Rehovat.

Bennet, R.J. & R.J. Chorley, 1978: *Environmental Systems*, London.

Berdoulay, V., 1978: The Vidal-Durkheim Debate, *Humanistic Geography*, eds. D. Ley & M.S. Samuels, Chicago, pp. 79/90.

Berry, B.J.L., 1964: Approaches to Regional Analysis: a Synthesis, *A.A.A.G.* 54, pp. 2/11.

Berry, R.J.L., 1968: A Synthesis of Formal and Functional Regions, Using a General Field Theory of Spatial Behavior, *Spatial Analysis*, eds. B.J.L. Berry & D.F. Merble, New York, pp. 419/428.

Berry, R.J.L., 1971: DIDO Analysis – GIGO or Pattern Recognition? *Perspectives in Geography* 1, North Ill. Univ., pp. 105/131.

Berry, R.J.L., 1973: A Paradigm for Modern Geography, *Directions in Geography*, ed. R.J. Chorley, London, pp. 3/21.

Berry, R.J.L., 1980: Creating Future Geography, *A.A.A.G.* 70, pp. 449/458.

Bertalanffy, v. L., 1951: An Outline of General Systems Theory, *Brit. Journ. Phil. Scie.*, pp. 134/165.

Bertalanffy, v. L., 1962: General Systems Theory; a Critical Review, *Gen. Syst.* 7, pp. 1/20.

Bertalanffy, v. L., 1969: Chance or Law, *The Alpbach Symposium 1968: Beyond Reductionism, New Perspectives in the Life Sciences*, eds. A. Koestler & J.R. Smythies, London, pp. 56/84.

Bertalanffy, v. L., 1971: *General Systems Theory*, Penguin, London.

Bertrand, G., 1984: Les géographes français et leur paysages, *Ann. de Géogr.*, pp. 218/229.

Birdsall, S.S. & J.N. Florin, 1978: *Regional Landscapes of the U.S. and Canada*, New York.

Birot, P., 1970: *Les régions naturelles du globe*, Paris.

Blache, de la P. Vidal, 1913: Des caractères distinctifs de la géographie, *Ann. de Géogr.*, pp. 289/299.

Blache, de la P. Vidal, 1926: *Principles of Human Geography*, London.

Blotenvogel, H.H., 1984: Zeitungsregionen in der Bundesrepublik Deutschland, *Erdkun.*, pp. 79/93.

Bode, H., F. Mostella, F. Turkey, C. Winsor, 1949: The Education of a Scientific Generalist, *Science* 109, p. 533.

Boguslaw, R., 1981: *The New 'Utopians'; a Study of System Design and Social Change*, New York.

Boudeville, J.R. [ed.], 1968a: *L'espace et les pôles de croissance*, Paris.

Boudeville, J.R., 1968b: Les notions d'espace et d'integration, *L'espace et les pôles de croissance*, ed. J.R. Boudeville, Paris, pp. 23/40.

Brassard, T. & J.C. Wieber, 1984: Le paysage; trois definitions, un model d'analyse et de cartographie, *L'esp. géogr.*, pp. 5/12.

Bratzel, P. & H. Muller, 1979: Regionalisierung der Erde, *Geogr. Rundsch.*, pp. 131/137.

Brookfield, H.C., 1964: Human Frontiers in Geography, *Econ. Geogr.*, pp. 283/303.

Brookfield, H.C., 1969: On the Environment as Perceived, *Progr. in Geogr.* 1, pp. 51/80.

Brunet, R., 1972a: Pour une théorie de la géographie régionale, *La pensée géographique française contemporaine*, Mélanges offerts à M.A. Meynier, St. Brieuc, pp. 649/662.

Brunet, R., 1972b: Les nouveaux aspects de la recherche géographique: rupture ou raffinement de la tradition, *L'esp. géogr.*, pp. 73/78.

Brunet, R., 1979: Systèmes et approches systémiques en géographie, *B.A.G.F.* No. 465, pp. 299/307.

Buchanan, K., 1977: Economic Growth and Cultural Liquidation: the Case of the Celtic Nation, *Radical Geography*, ed. R. Peet London, pp. 125/142.

Bunge, W.W., 1960: *Theoretical Geography*, PhD Dissertation, Univ. of Washington, Univ. Microfilms Inc. Ann Arbor.

Bunge, W.W., 1966: Locations are not Unique, *A.A.A.G.* 56, pp. 375/378.

Bunge, W.W., 1973: Ethics and Logic in Geography, *Directions in Geography*, ed. R.J. Chorley, London, pp. 317/331.

Bunting, T. & L. Guelke, 1979: Behavioral and Perception Geography; a Critical Apraisal, *A.A.A.G.* 69, pp. 448/462.

Burton, I., 1963: The Quantitative Revolution and Theoretical Geography, *Canad. Geogr.* 7, pp. 151/162.

Buttimer, A., 1971: *Society and Milieu in the French Geographical Tradition*, sixth in the monograph series, A.A.G., Chicago.

Buttimer, A., 1974: Values in Geography, Commission on College Geography, *A.A.A.G. Res. Pap. no. 24*, Washington DC.

Buttimer, A., 1976: Grasping the dynamism of Lifeworld, *A.A.A.G.* 66, pp. 277/292.

Buttimer, A., 1978: Charism and Context: the Challenge of 'la géographie humaine', *Humanistic Geography, Prospects and Problems*, eds. D. Ley & M.S. Samuels, Chicago, pp. 58/76.

Buttimer, A., 1983: *The Practice of Geography*, London.

Buttimer, A. & D. Seamon [eds.], 1980: *The Human Experience of Space and Place*◇, London.

Carson, R., 1962: *The Silent Spring*, N.Y..

Chapman, G.P.,1974: *Human and Environmental Systems; a Geographer's Appraisal*, London.

Chappell, J., 1981: Environmental Causation, *Themes in Geographical Thought*, eds. M.H. Harvey & B.P. Holly, London, pp. 163/186.

Chorley, R.J., 1962: Geomorphology and General System Theory, *U.S. Geolog. Surv. Prof. Pap.* No. 500 B.

Chorley, R.J. [ed.], 1973: *Directions in Geography*, London.

Chorley, R.J. & P. Haggett [eds.], 1965: *Frontiers in Geographical Teaching*, London.

Chorley, R.J. & P. Haggett [eds.], 1967: *Models in Geography*, London.

Chorley, R.J. & B.A. Kennedy, 1971: *Physical Geography – a System Approach*, London.

Chouinard, V., R. Fincher, M. Webber, 1984: Empirical Research in Scientific Human Geography, *Progr. in Hum. Geo.* 8, pp. 347/380.

Christaller, W., 1966: *Central Places in Southern Germany*, transl. W. Baskin, New Jersey.

Christensen, K., 1982: Geography as Human Science; a Philosophic Critique of the Positivist-Humanist Split, *A Search for Common Ground*, eds. P. Gould & G. Olsson, pp. 37/57.

Clarkson, J.D., 1968: The Cultural Ecology of a Chinese Village; Camerun Highlands, Malaysia, *Res. Pap. no. 114*, Dept. of Geogr. Univ. of Chicago.

Claval, P., 1979: *La nouvelle géographie*, Paris.

Claval, P., 1984a: The Concept of Social Space and the Nature of Social Geography, *New Zeal. Geogr.*, 40, pp. 105/109.

Claval, P., 1984b: Les languages de la géographie et le rôle du discours dans son évolution, *Ann. de Géogr.*, pp. 409/422.

Claval, P., 1985a: Causalité et géographie, *L'esp. geogr.*, pp. 109/115.

Claval, P., 1985b: Nouvelle géographie, communication et transparence, *Ann. de Géogr.*, pp. 129/144.

Claval, P., & E. Juillard, 1967: *Région et régionalisation dans la géographie française et dans d'autres sciences sociales*, Paris.

Cole, J.P. & C.A.M. King, 1968: *Quantitative Geography*, London.

Couclelis, H., 1982: Philosophy in the Construction of Geographic Reality, *A Search for Common Ground*, eds. P. Gould & G. Olsson, Chicago, pp. 195/138.

Couclelis, H. & R. Golledge, 1983: Analytic Research, Positivism and Behavioral Geography, *A.A.A.G.* 73, pp. 331/339.

Cowie, P.M., 1978: Geography, Value Laden Subject in Education, *Geogr. Educ.* 3, pp. 133/146.

Dainville, F., 1940: *La géographie des Humanistes*, Paris.

Darby, H.C., 1948: The Regional Geography of Thomas Hardy's Wessex, *Geogr. Rev*, pp. 426/443.

Daude, G., 1971: Essai de définition d'une géographie régionale dynamique, *Rev. de Géogr. de Lyon*, pp. 441/448.

Davis, W.M., 1899: The Geographical Cycle, *Geogr. Journ.*, pp. 481/504.

De Blij, H.J., S.S. Biresall, H.J. Stole, A.F. Ryan, 1974: *Essentials of Geography; Regions and Concepts*, New York.

Demek, J., 1978: The Landscape as a Geosystem, *Geoforum*, pp. 29/34.

Derruau, M., 1961: *Précis de géographie humaine*, Paris.

Dickinson, H.T., 1947: *City, Region and Regionalisation; a Geographical Contribution to Human Ecology*, London.

Dickinson, R.E., 1969: *The Makers of Modern Geography*, London.

Dickinson, R.E., 1970: *Regional Ecology; the Study of Man's Environment*, London.

Dilisio, G., 1983: *Maryland, a Geography*, Boulder CO.

Douglas, J.P., 1985: Literature and Geography, *Area* 17, pp. 117/122.

Drdoš, J., 1983: Landscape Research and its Anthropic Orientation, *Geo-Journ.* 7, pp. 155/160.

Dumolard, P., 1975: Région et régionalisation: une approche systémique, *L'esp. geogr.*, pp. 93/111.

Dumolard, P., 1980: Le concept de la région: ambiguités, paradoxes ou contradictions? *Trav. Inst. Geogr. Reims*, 41-42, pp. 21/32.

Dziewonski, K., 1967: Théorie de la région économique, *Mélanges offerts à M.O. Tulipe*, Gembloux, pp. 545/557.

Edwards, K.C., 1970: Regional Geography, *Geography, an Outline for the Intending Student*, ed. W.G. Balchin, London, pp. 99/114.

Entrikin, N.J., 1976: Contemporary Humanism in Geography, *A.A.A.G.* 66, pp. 615/632.

Ferrier, J.P., J.B. Racine, C. Raffestine, 1978: Vers un paradigme critique; matériaux pour un projet géographique, *L'esp. géogr.*, pp. 291/297.

Fielding, G.J., 1974: *Geography as Social Science*, New York.

Fien, J., 1981: Values Probing: on Integral Approach to Values Education in Geography, *Journ. of Geogr.*, 80, pp. 19/22.

Feyerabend, P., 1978: *Against Method; Outline of an Anarchistic Theory of Knowledge*, London.

Forde, C.D., 1939: Human Geography, History and Sociology, *Scot. Geogr. Magaz.* 55, pp. 217/235.

Frankl, V., 1969: [intervention in discussion, p. 219], *The Alpbach Symposium 1968: Beyond Reductionism, New Perspectives in the Life Sciences*, eds. A. Koestler & J.R. Smythies, London.

Frémont, A., 1976: *La région, espace vécu*, Paris.

Frémont, A., 1980: Espace vécu et la notion de la région, *Trav. Inst. Géogr. Reims*, 41-42, pp. 47/58.

French, H.M., & J.B. Racine [eds], 1971: *Quantitative and Qualitative Geography; la nécessite d'un dialogue*, Ottawa.

Gale, S, & G. Olsson [eds.] 1979: *Philosophy in Geography*, Dordrecht.

Galois, B., 1976: Ideology and the Idea of Nature; the Case of Peter Kropotkin, *Antipode* 8, No. 3, pp. 1/16.

Ganor, E., 1963: Oscillation of the Limit of Aridity in Israel, *Layaran* 13, pp. 136/142 [Hebrew].

Gatrell, A.C., 1983: *Distance and Space, a Geographical Perspective*, Oxford.

Geddes, P., 1915: *Cities in Evolution*, London [reprinted 1968].

George, P., 1981: La géographie, histoire profonde; à la recherche d'une notion globale de l'Espace, *Ann. de Géogr.*, pp. 203/210.

Giblin, B. [ed.], 1982: *Elisée Reclus, L'homme et la terre*, 2 vols., introd. et choix des textes par l'editeur, Paris.

Gibson, E.M.W., 1978: Understanding the Subjective Meaning of Places, *Humanistic Geography,Prospects and Problems*, eds. D. Ley & M.S. Samuels, Chicago, pp. 138/154.

Gibson, E.M.W., 1981: *Realism, Themes in Geographical Thought*, eds. M.E. Harvey & B.P. Holly, London, pp. 148/162.

Gilbert, E.W., 1960a: Geography and Regionalism, *Geography in the 20th Century*, ed. G. Taylor, pp. 345/371.

Gilbert, E.W., 1960b: The Idea of the Region, *Geography* 45, pp. 157/175.

Gilsen, E. & J. Nipper, 1984: Die Bedeutung von Innovation und Diffusion neuer Technologien fuer die Regionalpolitik, *Erdkun.*, pp. 205/215.

Glikson, A., 1955: *Regional Planning and Development*, Leiden.

Golledge, R., 1982: Fundamental Conflicts and the Search for Geographical Knowledge, *A Search for Common Ground*, eds. P. Gould & G. Olsson, Chicago, pp. 11/23.

Gottmann, J., 1955: *Virginia at mid-Century*, New York.

Gottmann, J., 1957: *Les marchées des matières premières*, Paris.

Gottmann, J., 1958: Regional Planning in France; a review, *Geogr. Rev.*, pp. 257/261.

Gottmann, J., 1961: *Megalopolis; the Urbanised Northeastern Seabord of the U.S.*, New York.

Gottmann, J., 1962: *A Geography of Europe,;*3d. ed., New York.

Gottmann, J., 1966: The Ethics of Living at Higher Densities, *Ekistics* 21, no. 123, pp. 141/145.

Gottmann, J., 1978: Urbanisation and Employment, towards a General Theory, *Town Plann. Rev.* 49, pp. 393/401..

Gottmann, J., 1980: Les frontières et les marchées; cloisonnement et dynamique du monde, *Geography and its Boundaries*, in Memory of Hans Boesch, ed. H. Kishimoto, Berne, pp. 53/58.

Gottmann, J., 1981a: Japan's Organization of space; Fluidity and Stability in a Changing Habitat, *Ekistics*, pp. 258/265.

Gottmann, J., 1981b: A Note on 'the Organisation of Space and the Freedom of the Individual' I.P.S.A. Research Comittee on Political Geography, typescript 2 pages.

Gottmann, J., 1984: Orbits, the Ancient Mediterranean Tradition of Urban Network, *The Twelfth J.S. Myres Memorial Lecture*, New College, Oxford, 3 May 1983, Leopard's Head Press, London..

Gould, P., 1979: Geography 1957-1977; the Augean Period, *A.A.A.G.* 69, pp. 139/151.

Gould, P., 1985: *Geographer at Work*, London.

Gould, P. & G. Olsson [eds.], 1982: *A Search for Common Ground*, London.

Gourou, P., 1936: *Les paysans du delta tonkinois*, étude de géographie humaine, Paris.

Gourou, P., 1947: *Les pays tropicaux*, Paris [4th ed., Paris 1966; engl. transl. The Tropical World, 1961, London].

Gourou, P., 1975: *Leçons de Géographie Humaine*, Paris.

Gradman, R., 1929: Dynamische Laenderkunde, *Geogr. Zeitschr.*, pp. 551 553.

Gradman, R., 1931: Das Laenderkundliche Schema, *Geogr. Zeitschr.*, pp. 540/548.

Gregory, D., 1978: *Ideology, Science and Human Geography*, London.

Grigg, D., 1967: Regions, Models and Classes, *Models in Geography*, eds. R.J. Chorley & P. Haggett, London, pp. 461/509.

Grossman, L., 1977: Man-Environment Relationship in Anthropology and Geography, *A.A.A.G.* 67, pp. 126/144.

Guelke, L., 1974: An Idealist Alternative in Human geography, *A.A.A.G.* 64, pp. 193/202.

Guelke, L., 1977: Regional Geography, *Prof. Geogr.* 29, pp. 1/7.

Geulke, L., 1981: Idealism, *Themes in Geographical Thought*, eds. M.E. Harvey & B.P. Holly, pp. 137/147.

Haase, G., & H. Richter, 1983: *Current Trends in Landscape Research, Geo-Journ.* 7, pp. 107/119.

Haegerstrand, T., 1973: The Domain of Human Geography, *Directions in Geography*, ed. R.J. Chorley, pp. 67/87.

Haegerstrand, T., 1976: Geography and the Study of Interaction between Nature and Society, *Geoforum*, pp. 329/334.

Haggett, P., A.D. Cliff, A. Frey [eds.], 1977: *Locational Analysis in Geography*, 2nd ed., London.

Haigh, M.J.,1985: Geography and General System Theory, Philosophical Homologies and Current Practice, *Geoforum*, pp. 191/203.

Hansen, N.M., 1968: *French Regional Planning*, Bloomington.

Harper, R.A. & T.H. Schmudde, 1984: *Between Two Worlds: an Introduction to Geography*, Dubuque.

Hart, J.F., 1982: The Highest Form of the Geographer's Art, *A.A.A.G.* 72, pp. 1/27.

Hartshorne, R., 1939: *The Nature of Geography. A Critical Survey of Current Thought in the Light of the Past*, Lancaster.

Hartshorne, R., 1955: 'Exceptionalism in Geography' re-examined, *A.A.A.G.* 45, pp. 205/244.

Hartshorne, R.,1958: The Concept of Geography as a Science of Space from Kant and Humboldt to Hettner, *A.A.A.G.* 48, pp. 97/108.

Hartshorne, R., 1959: *Perspective on the Nature of Geography*, Chicago.

Harvey, D., 1969: *Explanation in Geography*, London.

Harvey, D.,1979: Population, Resources and the Ideology of Science, *Philosophy in Geography*, eds. S. Gale & G. Olsson, pp. 155/185.

Harvey, M.E. & B.P. Holly [eds.], 1981a: *Themes in Geographical Thought*, London.

Harvey, M.E. & B.P. Holly, 1981b: Paradigm, Philosophy & Geographical Thought, *Themes in Geographical Thought*, eds. M.E. Harvey & B.P. Holly, London, pp. 11/17.

Helburn, N., 1982: Geography and the Quality of Life, *A.A.A.G.* 72, pp. 445/456.

Herbst, J., 1961: Social Darwinism and American Geography, *Proc. Amer. Phil. So.* 105, pp. 538/544/.

Hettner, A., 1932: Das Laenderkundliche Schema, *Geogr. Anzeig.* 33, pp. 1/6.

Hettner, A., 1934-1937: *Vergleichende Laenderkunde*, Leipzig.

Hill, M.R., 1981: Positivism: a 'Hidden' Philosophy in Geography, *Themes in Geographical Thought*, eds. M.E. Harvey & B.P. Holly, London, pp. 38/60.

House, J.W.,1970: Applied Geography, *Geography, an Outline for the Intending Student*, ed. W.G. Balchin, London, pp. 132/146.

Howard, E., 1902: *Garden Cities of Tomorrow*, London [reprinted 1946].

Huggett, R.J., 1976: A Schema for the Science of Geography, its Systems, Laws and Models, *Area*, pp. 25/30.

Huggett, R.J., 1980: *System Analysis in Geography*, Oxford.

Huggett, R.J., 1981: A Hard Line on Soft Systems, *Area*, pp. 224/226.

Huntington, E., 1924: *Civilisation and Climate*, 3d ed., New Haven.

Husserl, E., 1960: *Méditations cartésiennes* [engl. transl. by Dorion Chairns, *Cartesian Meditations*, The Hague].

Inbar, M., 1972: A Geomorphic Analysis of a Catastrophic Flood in a Mediterranean Basaltic Watershed, Dept. of Geogr., Univ. of Haifa, typescript.

Isard, W.,1960: Methods of Regional Analyzis; *Introduction to Regional Sciences*, New York.

Isnard, H., 1978(a): *L'èspace géographique*, coll. 'Le Géographe' PUF. Paris.

Isnard, H., 1978(b): Pour une géographie 'empiriste', *Ann. de Géogr.*, pp. 513/517.

Isnard, H., 1980: Méthodologie et géographie, *Ann. de Géogr.*, pp. 129/143.

Isnard, H., 1985: Espace et temps en géographie, *Ann. de Géogr.*, pp. 535/545.

Isnard, H., J.B. Racine, H. Reymond, 1981: *Problématique de la géographie*, Paris.
Ivanička, K., 1980: *Prognóza ekonomicko-geografických systémov*, Bratislava.
Jackson, J.B., 1972: Metamorphosis, *A.A.A.G.* 62, pp. 155/158.
Jackson, R.H.& A. Sofer, 1986: Irrigation in Sub-Humid Environments; a Comparison of Three Cultures, *Geoforum* 17, pp. 383/401.
James, P.E., 1952: Towards a Further Understanding of the Regional Concept, *A.A.A.G.* 42, pp. 195/222.
James, P.E. [ed.], 1959a: *New Viewpoints in Geography*, Nat. Coun. for Soc. Studies, Washington DC.
James, P.E., 1959b: American Geography at Mid-Century, *New Viewpoints in Geography*, ed. P.E. James, Washington DC, pp. 10/18.
James, P.E., 1967: On the Origin and Persistance of Error in Geography, *A.A.A.G.* 57, pp. 1/24.
James, P.E., 1972: *All Possible Worlds; a History of Geographical Ideas*, Indianapolis.
James, P.E. & C.F. Jones, 1954: *American Geography; Inventory and Prospect*, Syracuse.
Jeans, D.N., 1974: Changing Formulation of the Man-Environment Relationship in Anglo-American Geography, *Journ. of Geogr.*, 73, pp. 36/40.
Johnston, R.J., 1968: Choice in Classification: the Subjectivity of Objective Methods, *A.A.A.G.* 58, pp. 575/589.
Johnston, R.J., 1983: *Geography and Geographers*, London, 2nd ed..
Johnston, R.J., 1985: The World is our Oyster, *Geogr. Futures*, ed. R. King, Sheffield, pp. 112/128.
Johnston, R.J., 1986: Four Fixations and the Quest for Unity in Geography, *Trans. Inst. Brit. Geogr.* New Ser. 11, pp. 449/453.
Jones, E., 1980: Social Geography, *Geography, Yesterday and Tomorrow*, ed. E.H. Brown, Oxford, pp. 251/262.
Jones, E., 1984: On the specific Nature of Space, *Geoforum*, pp. 5/9.
Jones, E., 1985: The Paradigm cycle, *Area* 17, pp. 169.
Juillard, E., 1962: La région; essai de définition, *Ann. de Géogr.*,   pp. 483/499.
Juillard, E., 1974: *La région, contribution a une géographie générale des espaces régionaux*, Paris.
Karmon, Y., 1959: *The Sharon*, PhD Thesis, Hebrew University, Jerusalem [Hebrew, typescript].
Kimble, G.H.T., 1951: The Inadequacy of the Regional Concept, *London Essays in Geography*, eds. L.D. Stamp & S.W. Wooldridge, London, pp. 151/174.
King, R. [ed.], 1985: *Geographical Futures*, Sheffield.
Knight, D.B., 1982: Identity and Territory; Geographical Perspectives on Nationalism and Regionalism, *A.A.A.G.* vol. 72, pp. 514/531.
Koestler, A., 1959: *The Sleepwalkers, a History of Man's Changing Vision of the Universe*, London.
Koestler, A., 1964: *The Art of Creation*, London.
Koestler, A., 1969: Beyond Atomism and Holism – the concept of the Holon, *The Alpbach Symposium 1968: Beyond Reductionism, New Perpectives in the Life Sciences*, eds. A. Koestler and J.R. Smythies, London, pp. 192/232.
Koestler, A. & J.R. Smythies [eds.], 1969: *The Alpbach Symposium 1968: Beyond Reductionism, New Perspectives in the Life Sciences*, London.
Kohn, C.F., 1970: The 1960's: a Decade of Progress in Geographical Research and Instruction, *A.A.A.G.* 60, pp. 211/219.
Kropotkin, P., 1902: *Mutual Aid: a Factor in Evolution*, London.
Kuklinski, A., 1985: Four Interpretations of Regional Development, a note for Discussion, *Moncton Regional Development Forum*, typed letter, May 13, 1985.
Labasse, J., 1955: *Les capitaux et la région*, Lyon.
Lacoste, Y., 1976: *La géographie, ca sert, d'abord, à faire la guerre*, Paris.
Laity, A.L., 1984: Perceiving Regions as Scattered Objects, *Prof. Geogr.*, pp. 285/292.
Langton, J., 1972: Potentialities and Problems of Adopting a Systems' Approach to the Study of Change in Human Geography, *Progr. in Geogr.* 4, pp. 125/179.
Langton, J., 1984: The Industrial Revolution and the Regional Geography of England, *Trans. Inst. Brit. Geogr.* New ser., 9, pp. 145/167.
Lauer, Q., 1965: *Phenomenology – its Genesis and Prospect*, New York.
Lavrov, S.B. & G.V. Sdasyvk, 1984: The Evolution of Regional Development Concepts; some New Trends, *Geoforum*, pp. 11/17.

Le Berre, M., 1980: Heur et Malheur de la géographie régionale, *Trav. Inst. Géogr. Reims*, No. 41-42, pp. 3/19.

Leghausen, P., 1974: Ecology, Behaviour, Quality of Life and the Method of Quantification, *System Approaches and Environmental Problems*, ed. H.W. Gottinger, Goettingen, pp. 335/349.

Leighly, J. [ed.], 1963: *Land and Life, a Selection from the Writing of C.O.Sauer*, Univ. of California Press, Berkeley and L.A.

Levins, R. & R. Lewontin, 1984: *The Dialectic Biologist*, Cambridge. Mass..

Lewis, P.E., 1979: Axioms for Reading the Landscape, *The Interpretation of Ordinary Landscapes*, ed. D.B. Meinig, pp. 11/32.

Ley, D., 1980: Geography without Man; a Humanistic Critic, *School of Geogr., Univ. of Oxford Research Pap.* 24, 25 pp..

Ley, D., 1983: Cultural-Humanistic Geography, *Progr. in Hum. Geogr.* 7, pp. 267/275.

Ley, D. & M.S. Samuels [eds.], 1978a: *Humanistic Geography, Prospects and Problems*, Chicago.

Ley, D. & M.S. Samuels, 1978b: Introduction: Contex of Modern Humanism in Geography, *Humanistic Geography, Prospects and Problems*, eds. D. Ley & M.S. Samuels, Chicago, pp. 1/17.

Lichtenberger, E., 1979: The Impact of Political system upon Geography: the case of FDR and DDR, *Prof. Geogr.*, pp. 201/211.

Loesch, A., 1954: *The Economics of Location*, transl. by W.H. Woglom, New Haven.

Loewenthal, D.,1961: Geography, Experience and Imagination; towards a Geographical Epistemology, *A.A.A.G.* 51, pp. 241/260.

Loewenthal, D. & H.C. Prince, 1964: The English Landscape, *Geogr. Rev.* 54, pp. 309/346.

Loewenthal, D. & H.C. Prince, 1965: English Landscape Tastes, *Geogr. Rev.* 55, pp. 186/222.

Loewenthal, D. & M.J. Bowden, 1976: *Geographies of the Mind*, N.Y.

Lutwack, L., 1984: *The Role of Place in Literature*, Syracuse Univ. Press..

Marchand, B., 1974: Quantitative Geography: Revolution or counter-Revolution? *Geoforum* 17, pp. 15/23.

Marchand, B., 1978: A Dialectic Approach in Geography, *Geogr. Anal.*, 10, pp. 105/119.

Marchand, B.,1979: Dialectics and Geography, *Philosophy in Geography*, eds. S. Gale & G. Olsson, Dordrecht, pp. 237/267.

Mackinder, M.P., 1919: *Democratic Ideals and Reality – a Study in the Politics of Reconstruction*, London.

Mackintosh, J.P., 1968: *The Devolution of Power: Local Democracy, Regionalism and Nationalism*, London.

Mallony, W.E. and P. Simpson-Hously [eds.] 1987: *Geography & Literature: A meeting of the Disciplines*. Syracuse Univ. Press..

Martin, J.P. & H. Nonn, 1980: La notion d'intégration régionale, *Trav. Inst. Géogr. Reims*, No. 41-42, pp. 33/46.

Martonne, de, E., 1902: *La Valachie*, Paris.

Mazúr, E., 1983: Landscape Synthesis – Objectives and Tasks, *Geo. Journ.* vol. 7, pp. 101/106.

Mazúr, E. & J.Urbanek, 1983: Space in Geography, *Geo.Journ.* 7, pp. 139/143.

McDonald, J.R., 1966: The Region, its Conception, Design and Limitations, *A.A.A.G.* 56, pp. 510/528.

Mead, W.R., 1980: Regional Geography, *Geography, Yesterday and Tomorrow*, ed. E.H. Brown, Oxford, pp. 292/302.

Meinig, D.W. [ed.], 1979a: *The Interpretation of Ordinary Landscapes*, New York.

Meinig, D.W., 1979b: The Beholding eye, *The Interpretation of Ordinary Landscapes*, ed. D.W. Meinig, pp. 33/48.

Meinig, D.W., 1983: Geography as an Art, *Trans. Inst. Brit. Geogr.*, New Series 8, pp. 314/328.

Mensching, H., 1986: Is the Desert Spreading? Desertification in the Sahel Zone of Africa, *Appl. Geogr. and Develop.*, 27, Tuebingen, pp. 7/18.

Miernyk, W.H., 1982: *Regional Analysis and Regional Policy*, Cambridge, Mass..

Mikesell, M.W., 1972: *Landscape, Man, Space and Environment*, eds. P.W. English & R.C. Mayfield, New York.

Mikesell, M.W., 1978a: The rise and decline of 'Sequent Occupance'; a Chapter in the History of

American Geography, *Geographies of the Mind*, eds. D. Loewenthal & M.J. Bowden, New York, pp. 149/169.

Mikesell, M.W., 1978b: Tradition and Innovation in Cultural Geography, *A.A.A.G.* 68, pp. 1/16.

Miller, J.G., 1978: *Living Systems*, New York.

Minshull, R., 1962: *Regional Geography, Theory and Practice*, London.

Morgan, R., 1981: System Analysis – a Problem of Methodology? *Area* 13, pp. 219/223; 227/229.

Morrill, R.L., 1983: The Nature, Unity and Value of Geography, *Prof. Geogr.* 35, pp. 1/9.

Morrill, R.L., 1984: The Responsability of Geography, *A.A.A.G.* 74, pp. 1/8.

Moss, R.P., 1979: On Geography as Science, *Geoforum* 10, pp. 223/233.

Muhsam, H.U., 1955: Enumerating the Beduin of Palestine, *Scripta Hierosolymitana* III, pp. 265/280.

Neef, E., 1967: *Die theoretischen Grundlagen der Landschaftslehre*, Leipzig.

Nicolas, O.G., 1984: *L'espace originel. Axiomatisation de la Géographie*, Berne.

Nir, D.,1968: *La Vallée de Beth Shean, une région a la lisière du desert*, Paris.

Nir, D., 1974: *Methods in Regional Geography*, [Hebrew], Jerusalem.

Nir, D., 1983: *Man, a Geomorphological Agent; an Introduction to Anthropogeomorphology*, Jerusalem and Dordrecht.

Nir, D., 1985: La valeur socio-culturelle de la géographie régionale, *L'Esp. géogr.*, pp. 69/71.

Nir, D., 1986a: Peter Kropotkin and the Israeli Utopia [Hebrew], *Mehkarim* [Stud. in the Geogr. of Isr.] vol. 12, pp. 11/20.

Nir, D., 1986b: Coastal Regions as Interdependant Socio-Natural Systems, *Cities on the Sea, Past and Present*, The First Inter. Symp. on Harbours, Port Cities and Coastal Topography, Haifa, September 22-29, ed. A. Raban,.pp. 13/16.

Nir.D., 1987: The Terms 'Region' and 'Landscape' considered from the Systems' Approach, *Geoforum*, Vol. 18, no. 2, pp. 187-202.

Norton, W., 1984: The Meaning of Culture in Cultural Geography, *Journ. of Geogr.*, pp. 145/148.

Olsson, G., 1975: Birds in Egg, *Michigan Geogr. Publ*, No. 15, Ann Harbor.

Olsson, G., 1978: Far Cries from a Memoralising Mamafesta, *Humanistic Geography, Prospects and Problems*, eds. D. Ley & M.S. Samuels, pp. 109/120.

Olsson, G., 1979: Social Science and Human Action, or Hitting your Head against the Ceiling of Language, *Philosophy in Geography*, eds. S. Gale & G. Olsson, Dordrecht, pp. 287/307.

Paasen, v. Christian, 1981: The Philosophy of Geography; from Vidal to Haegerstand, *Space and Time in Geography*, ed. A. Pred, *Lund Stud. in Geogr.*, ser. B. Human geogr. No. 48, pp. 17/29.

Parsons, J.J., 1977: Geography as Exploration and Discovery, *A.A.A.G.* 67, pp. 1/16.

Parsons, J.J., 1985: On 'Bioregionalism' and 'Watershed Consciousness', *Prof. Geogr.* 37, pp. 1/6.

Paterson, J.H., 1975: Writing Regional Geography; Problems and Progress in the Anglo-American Realm, *Progr. in Geogr.* 6, pp. 1/26.

Peet, R. [ed.], 1977: *Radical Geography; Alternative Viewpoint on Contemporary Social Issues*, London.

Perroux, F., 1968a: Les espaces économiques, *L'espace et les pôles de croissance*, ed. J.R. Boudeville, Paris, pp. 5/22.

Perroux, F., 1968b: La construction analytique de la région, *L'espace et les pôles de croissance*, ed. J.R. Boudeville, Paris, pp. 63/84.

Philbrick, A.K., 1957: Principles of Areal Functional Organisation in Regional Human Geography, *Econ. Geogr.* 33, p. 300/336.

Philbrick, A.K., 1960: Practical Suggestion for the more Systematic Construction of Regional Generalizations, *A.A.A.G.* 50, pp. 340/.

Pickles, J., 1985: *Phenomenology, Science and Geography*, Cambridge.

Pivetau, J.L., 1969: Le sentiment d'appartenance régionale en Suisse, *Rev. de Géogr. Alp.* 59, pp. 364/386.

Platt, R.S.,1935: Field Approach to Regions, *A.A.A.G.* 25, pp. 153/174.

Pred, A. [ed.], 1981: Space and Time in Geography, Essays Dedicated to T. Haegerstrand, *Lund Stud. in Geogr. series B. Human Geography*, 48.

Pred, A., 1984: Place as Historically Contingent Process: Structuralism and the Time-Geographyof becoming Places, *A.A.A.G.* 74, pp. 279/297.

Preobrazhenskiy, V.S., 1983: Geosystem as an Object of Landscape study, *Geo-Journ.*, pp. 131/134.

Quaimi, M., 1982: *Geography and Marxism*, Oxford.

Ratzel, F., 1882: *Anthropogeographie, oder Grundzuege der Anwendung der Erdkunde auf die Geschichte*, Stuttgart.

Read, H., 1968: *The Cult of Sincerity*, New York.

Relph, E., 1977: Humanism, Phenomenology and Geography, *A.A.A.G.* 67, pp. 177/183.

Relph, E., 1981a: *Rational Landscapes and Humanistic Geography*, London.

Relph, E., 1981b: Phenomenology, *Themes in Geographical Thought*, eds. M.E. Harvey & B.P. Holly, London, pp. 99/114.

Renner, G.T., 1935: The Statistical Approach to Region, *A.A.A.G.* 25, pp. 137/152.

Reynaud, A., 1974: La géographie entre le mythe et la science, essai d'epistémologie, *Trav. Inst. Géogr. Reims* 18-19.

Reynaud, A., 1982: La Géographie, science sociale, *Trav. Inst. Géogr. Reims* 49-50.

Richmond, G.M., 1969: Stratigraphie comparée des terrains quaternaires des Alps et des Montagnes Rocheuses, *Etude sur le Quaternaire dans le Monde*, VIII Congr. INQUA, pp. 7/25.

Robinson, G.W.S., 1953: The Geographical Region; Form and Function, *Scott. Geogr. Mag.*, 69, pp. 49/59.

Rougerie, G., 1977: *Géographie des Paysages*, Paris.

Russell, D., 1979: An Open Letter on the Dematerialization of the Geographical Object, *Philosophy in Geography*, eds. S. Gale & G. Olsson, Dordrecht, pp. 329/344.

Russell, J.A., A.W. Booth, S. Poole, 1954: Military Geography, *American Geography: Inventory and Prospect*, eds. P.E. James & C.F. Jones, pp. 485/495.

Sanguin, A.L., 1981: La géographie humaniste ou l'approche phénoménologique des lieux, des paysages et des espaces, *Ann. de Géogr.*, pp. 560/587.

Sanguin, A.L., 1984: Le paysage politique; quelques considerations sur un concept resurgent, *L'Esp. geogr.*, pp. 23/31.

Saarinen, T.E., 1976: *Environmental Planning; Perception and Behavior*, Boston.

Samuels, M.S.,1969: *Science and Geography: an Existential Appraisal*, PhD Dissertation, Univ. of Washington.

Samuels, M.S., 1979: The Biography of Landscape, *The Interpretation of the Ordinary Landscape*, ed. D.W. Meinig, pp. 51/88.

Samuels, M.S., 1981: On Existential Geography, *Themes in Geographical Thought*, eds. M.E. Harvey & B.P. Holly, pp. 115/132.

Sauer, C.O., 1924: The Morphology of Landscape, *Univ. of Cal. Publ. in Geogr.* 2, No 2, pp. 19/54.

Schaefer, F.K., 1953: Exceptionalism in Geography; a Methodological Examination, *A.A.A.G.* 43, pp. 226/249.

Schattner, I., 1962: The Lower Jordan Valley, *Scripta Hierosolymitana*, vol. XI, Jerusalem.

Schumacher, E.F., 1975: *Small is Beautiful; Economics as if People Mattered*, New York.

Scolnikov, S., 1984: An Image of Perfection, Nature and Reason, in Plato's Timaeus, *IYYUN Phil. Quaterly* [Hebrew] 30, nos. 3-4, pp. 206/227.

Shmueli.A., 1970: *Settlement of the Beduin in the Judean Desert* [Hebrew], Tel Aviv.

Sholem, G., 1982: *From Berlin to Jerusalem* [Hebrew], Tel Aviv.

Sitwell, O.F.G. & G.R. Latham, 1979: Behavioural Geography and the Natural Landscape, *Geogr. Annaler* series B 61, pp. 51/63.

Sorre, M., 1948: La notion de genre de vie et sa valeur actuelle, *Ann. de Géogr.* LVII, pp. 97/108; 193/204.

Spethmann, H., 1927: Neue Wege in der Laenderkunde, *Zeit. f. Geopol.*, pp. 989/998.

Spethmann, H., 1928: *Dynamische Laenderkunde*, Breslau.

Spethmann, H., 1931: *Das laenderkundliche Schema in der deutschen Geographie. Kaempfe fuer Fortschritt und Freiheit*, Berlin.

Spielberg, H., 1974; Phenomenological Movement, *Enc. Brit.*

Stewig, A. [ed.], 1979: *Probleme der Laenderkunde*, Darmstadt.

Stoddart, D.R., 1965: Geography and the Ecological Approach: the Ecosystem as a Principle and Method, *Geogr.* 50, pp. 242/251.

Stoddart, D.R., 1966: Darwin's Impact on Geography, *A.A.A.G.* 56, pp. 683/698.

Szymanski, R. & J.A. Agnew, 1981: Order and Scepticizm: Human Geography and the Dialectic of Science, *A.A.G. Res. Publ. in Geogr.*, 79 pp., Washington DC.

Taaffe, E.J.,1985: Comments on Regional Geography, *Journ. of Geogr.*, pp. 96/97.

Talbot, W.J., 1961: Land Utilization in the Arid Region of Southern Africa, *Arid Zone Res.* 17, pp. 299/331.

Tatham, G., 1951: Geography in the 19th Century, *Geography in the Twentieth Century*, ed. G. Taylor, New York, pp. 69/79.

Thuenen, v. J.H., 1966: *The Isolated State*, Oxford.

Topchiev, A.G., 1984: The Concept of Territorial Structure, *Sov. Geogr.* XXV, pp. 656/664.

Tornquist, G., 1979: On Fragmentation and Coherence in Regional Research, *Lund Stud. in Geogr.* series B No. 45, pp. 1/40.

Tornquist, G., 1981: On Arenas and Systems, *Space and Time in Geography*, Essays dedicated to T. Haegerstand, ed. A. Pred, *Lund Stud. in Geogr.* Series B 48, pp. 109/120.

Tricart, J. & M. Rochefort, 1953: *Initiation aux Travaux Pratiques*, Paris.

Tuan, Yi-Fu, 1975: Space and Place: Humanistic Perspective, *Progr. in Geogr.* 6, pp. 211/251.

Tuan, Yi-Fu, 1976: Geopiety – a Theme in Mass Attachment to Nature and Place, *Geography of the Mind*, eds. D. Loewenthal & M.J. Bowden, New York, pp. 3/39.

Tuan, Yi-Fu, 1978: Literature and Geography; Implications for the Geographical Research, *Humanistic Geography, Prospects and Problems*, eds. D. Ley & M.S. Samuels, Chicago, pp. 194/200.

Turnock, D., 1967: The Region in Modern Geography, *Geogr.* 52,pp. 374/383.

Uberoi, J.P.S., 1978: *Science and Culture*, Bombay.

Ullman, E.L, 1953: Human Geography and Area Research, *A.A.A.G.* 43, pp. 54/66.

Ullman, E.L., 1954: Regional Structure and Arrangement [reprinted in *Geography as Spatial Interaction* 1980, pp. 67/79].

Unstead, J.F.,1935: A System of Regional Geography, *Geogr.*, pp. 175/187.

Watson, M.K., 1978: The Scale Problem in Human Geography, *Geogr. Annal.* series B, pp. 36/47.

Weiss, P.A., 1969: The Living System: Determinism Stratified, *The Alpbach Symposium 1968: Beyond Reductionism, New Perspectives in the Life Sciences*, eds. A. Koestler & J.R. Smythies, London, pp. 3/55.

Whittlesey, D., 1954: The Regional Concept and the Regional Method, *American Geography, Inventory and Prospects*, eds. P.E. James & C.F. Jones, pp. 21/68.

Wilson, A.G., 1980: Theory in Human Geography: a Review Essay, *Geography, Yesterday and Tomorrow*, ed. H.E. Brown, London, pp. 201/215.

Wilson, A.G., 1981: *Geography and the Environment*, New York.

Wirth, E., 1984: Geographie als moderne Theorieorientierte Sozial-wissenschaft, *Erdkun.*, pp. 73/79.

Wright, J.K., 1966a: *Human Nature in Geography*, Harvard.

Wright, J.K., 1966b: Terrae Incognitae, *Human Nature in Geography*, by J.K. Wright, pp. 68/88.

Wrigley, E.A., 1965: Changes in the Philosophy of Geography, *Frontiers in Geographical Teaching*, eds. R.J. Chorley & P. ëHaggett, London, pp. 3/20.

Yair, A., 1983: Hillslope Hydrology, Water Harvesting and Areal Distributions of some Ancient Agricultural Systems in the Negev [Israel], *Journ. Arid Envir.* 6[3], pp. 283/301.

Zadeh, L.A., 1974: On the Analysis of Large-Scale Systems, *System Approaches and Environmental Problems*, ed. H.W. Gottinger, Goetingen, pp. 23/37.

Zobler, C., 1957: Statistical Testing of Regional Boundaries, *A.A.A.G.* 47, pp. 83/95.

Zonneveld, J.I.S., 1983; Some Basic notions in Geographical Synthesis, *Geo-Journ.* 7, pp. 121/129.

INDEX

The GeoJournal Library

1. B. Currey and G. Hugo (eds.): *Famine as Geographical Phenomenon.* 1984
   ISBN 90–277–1762–1
2. S. H. U. Bowie, F.R.S. and I. Thornton (eds.): *Environmental Geochemistry and Health.* Report of the Royal Society's British National Committee for Problems of the Environment. 1985   ISBN 90–277–1879–2
3. L. A. Kosiński and K. M. Elahi (eds.): *Population Redistribution and Development in South Asia.* 1985   ISBN 90–277–1938–1
4. Y. Gradus (ed.): *Desert Development.* Man and Technology in Sparselands. 1985   ISBN 90–277–2043–6
5. F. J. Calzonetti and B. D. Solomon (eds.): *Geographical Dimensions of Energy.* 1985   ISBN 90–277–2061–4
6. J. Lundqvist, U. Lohm and M. Falkenmark (eds.): *Strategies for River Basin Management.* Environmental Integration of Land and Water in River Basin. 1985   ISBN 90–277–2111–4
7. A. Rogers and F. J. Willekens (eds.): *Migration and Settlement.* A Multiregional Comparative Study. 1986   ISBN 90–277–2119–X
8. R. Laulajainen: *Spatial Strategies in Retailing.* 1987   ISBN 90–277–2595–0
9. T. H. Lee, H. R. Linden, D. A. Dreyfus and T. Vasko (eds.): *The Methane Age.* 1988   ISBN 90–277–2745–7
10. H. J. Walker (ed.): *Artificial Structures and Shorelines.* 1988
    ISBN 90–277–2746–5
11. A. Kellerman: *Time, Space, and Society.* Geographical Societal Perspectives. 1989   ISBN 0–7923–0123–4
12. P. Fabbri (ed.): *Recreational Uses of Coastal Areas.* A Research Project of the Commission on the Coastal Environment, International Geographical Union. 1990   ISBN 0–7923–0279–6
13. L. M. Brush, M. G. Wolman and Huang Bing-Wei (eds.): *Taming the Yellow River: Silt and Floods.* Proceedings of a Bilateral Seminar on Problems in the Lower Reaches of the Yellow River, China. 1989   ISBN 0–7923–0416–0
14. J. Stillwell and H. J. Scholten (eds.): *Contemporary Research in Population Geography.* A Comparison of the United Kingdom and the Netherlands. 1990
    ISBN 0–7923–0431–4
15. M. S. Kenzer (ed.): *Applied Geography.* Issues, Questions, and Concerns. 1989   ISBN 0–7923–0438–1
16. D. Nir: *Region as a Socio-environmental System.* An Introduction to a Systemic Regional Geography. 1990   ISBN 0–7923–0516–7